CHICAGO IN STONE AND CLAY

CHICAGO IN STONE AND CLAY

A Guide to the Windy City's Architectural Geology

Raymond Wiggers

NORTHERN ILLINOIS UNIVERSITY PRESS

AN IMPRINT OF CORNELL UNIVERSITY PRESS   ITHACA AND LONDON

First published 2022 by Cornell University Press
Printed in the United States of America

Library of Congress Cataloging-in-Publication Data

Names: Wiggers, Ray, author.
Title: Chicago in stone and clay : an exploration in architectural geology / Raymond Wiggers.
Description: Ithaca, [New York] : Northern Illinois University Press, an imprint of Cornell University Press, 2022. | Includes bibliographical references and index.
Identifiers: LCCN 2021051822 (print) | LCCN 2021051823 (ebook) | ISBN 9781501765063 (paperback) | ISBN 9781501765070 (pdf) | ISBN 9781501765087 (epub)
Subjects: LCSH: Building materials—Illinois—Chicago—Guidebooks. | Stone buildings—Illinois—Chicago—Guidebooks. | Building stones—Illinois—Chicago—Guidebooks. | Architecture—Illinois—Chicago—Guidebooks. | Geology—Illinois—Chicago—Guidebooks. | Chicago (Ill.)—Buildings, structures, etc.—Guidebooks.
Classification: LCC TA425 .W54 2022 (print) | LCC TA425 (ebook) | DDC 624.1/830977311—dc23/eng/20220217
LC record available at https://lccn.loc.gov/2021051822
LC ebook record available at https://lccn.loc.gov/2021051823

To the memory of my mother,
Irene Mary Wiggers
You taught me language
(the reason for a blessing, not a curse)

That astonishing Chicago—a city where they are always rubbing the lamp, and fetching up the genii, and contriving and achieving new impossibilities. It is hopeless for the occasional visitor to try to keep up with Chicago—she outgrows his prophecies faster than he can make them. She is always a novelty; for she is never the Chicago you saw when you passed through last time.

—Mark Twain, *Life on the Mississippi*

It is the old story of Eyes and No Eyes. What to the used-up man is an insufferably dismal trudge may become to the enthusiast a regular sporting tour; and I shall be more than satisfied with my work if I can awaken in any visitor . . . the same interest in an otherwise unattractive building that a knowledge of wild flowers, or birds, or insects, imparts to an otherwise stupid country walk in spring.

—Rev. H. W. Pullen, *Handbook of Ancient Roman Marbles*, 1894

Contents

Maps and Figures

Maps

Figures

Acknowledgments

To put it squarely, no other book project of mine has benefited half so much from the expertise of others. I'm more in hock to a larger number of persons from many walks of life than ever before. And I find that this is a wonderful place for an author to be.

First thanks must certainly go to NIU Press senior acquisitions editor Amy Farranto, who showed early and sustained interest in the ever-developing concept of this guide, provided sustained and crucial help in shaping the book into its current form, and guided me patiently and skillfully, as only a talented and experienced editor can, through the rocks and shoals of the review and manuscript-submission processes. A special nod of gratitude is also due to Cornell University Press acquisitions assistant Jacqulyn Teoh, who graciously dealt with the technical aspects of this book's proper submission and approval. The same applies to the production phase, most ably overseen by CUP assistant managing editor Karen Laun, with insightful copyediting by Ellen Douglas.

And this book would have just remained one of those interesting, unfulfilled ideas we all have had I failed to learn, and learn a lot, from the geologists and other stone experts who shared their expertise and access to seminal publications, recently published mapping, and other resources. Their number includes the ever-helpful and ever-knowledgeable John Nelson, now retired from the Illinois State Geological Survey, who provided excellent in-depth insight into the Pennsylvanian-age building stones of our state and Iowa; Todd Thompson, director of the Indiana Geological & Water Survey, who suggested a top-notch source on the famous Salem Limestone produced in his state; archivist Jennifer Lanman, also of the ISGWS, who oversaw the production of one of this book's maps; and Patrick Mulvaney, of the Missouri Geological Survey, who guided me to a treasure trove of geospatial data and publications on his state's rich mining and building-stone legacy. The Norwegian Geological Survey's Tom Heldal did an excellent job explaining the rare and previously enigmatic Støren Trondhjemite found on Chicago's 1 Prudential Plaza; and I greatly benefited from Aberdeen University professor Gordon Walkden's vast knowledge of Scottish granites and British building stones in general. Massachusetts state geologist Stephen Mabee took the time to check through the literature to confirm the revised radiometric date of the Bay State's famous Quincy Granite. In Pennsylvania, Joseph Jenkins, author and walking encyclopedia on slates and slate roofing, confirmed some of my own identifications and gave me additional information. I also appreciate the efforts of the Iowa Geological Survey's Ryan Clark; Peter Lemiszki, chief geologist of the Tennessee Survey; and Bill Prior at the Arkansas Geological Survey.

Of all the architects, historians, park officials, and landmark staff members contacted, several deserve signal recognition. Jim Havlat, of the Forest Preserve District of Cook County, taught me much about the 1930s Camp Sag Forest quarry and the design and construction history of Chicagoland's stone park structures. And at the Camp Sag Forest property itself—now the Sagawau Environmental Learning Center—Naturalist Negin Almassi gave me an impromptu midwinter tour of her site, and, with her boss Director Mike Konrath, supplied me with much additional information on Civilian Conservation Corps history, forest-preserve architecture, and Sagawau's exposed Silurian reef. Separately, Glessner House curator Bill Tyre repeatedly shared insights into the history and building materials of both the Prairie Avenue and Pullman neighborhoods; and Nate Lielasus not only provided insights into the design and construction of the Second Presbyterian Church but also graciously shared with me his excellent survey of Chicago's West Side Artesian quarries and buildings that feature their distinctive, bitumen-spotted stone. My thanks also to Haylie Carlson at the firm of Hammond Beeby and Babka, who furnished plans for the Harold Washington Library Center.

I suspect I could write a separate book just on what I learned from stone and brick suppliers, and from terra-cotta and ceramic producers and restorationists. From company presidents and their sales reps to veteran quarrymen, these individuals were so consistently helpful and well-versed in their products that I'm tempted to slather on far too many superlatives to describe them. Quite a few also generously contributed to the small mountain range of samples and reference specimens that now takes up most of the interior volume of my home. Their roster includes Jerry Vinci, executive vice president of the Illinois Brick Company; the eminently helpful Sheena Owen of Toronto's North Country Slate; Rich Carreras of Missouri's Earthworks Company; Polycor's Sylvie Beaudoin; Cleveland Quarries president Zachary Carpenter; Daryl Petit of Connecticut's Stony Creek Quarry; Lydia Lascola of Boston Valley Terra Cotta; Eric Dent of Chicago's Colonial Brick Company; Coldspring Corporation's marketing coordinator Stacy Gregory; Au Sable Forks, New York quarry general manager Rick Barber; President Andrew Baird and Mary Neely, both of Phenix Marble; Kimberley Olson at New Hampshire's Fitzwilliam Granite Company; and Tennessee Marble Company's Josh Buchanan. No less generous were George Hibben of Cincinnati's famous Rookwood Pottery; Mary Cloutier at Fletcher Granite; Doug Norman of North Carolina's Mount Airy Granite; Sue Lockwood at Dakota Granite; Marco Pezzica at Colorado Stone Quarries; Peter Prvulovic at Vermont

Quarries Corporation; Ed Smith of Williams Stone; Scott Balliew at Wisconsin's Eden Valders Stone; Kevin Aune of Ohio's Briar Hill Stone; and the staff of Keystone Memorials in Elberton, Georgia.

Managers and staff of a number of buildings and other facilities described in this book were also contacted whenever possible, and one, Jennifer Cervantes at Graceland Cemetery, deserves a separate and hearty tip of the hat for her consistent responsiveness and hard work. She patiently fielded a whole litany of questions over the course of many months, ferreted through office records, unearthed maps and documents, and even conducted site checks for me.

While their efforts and expertise are often woefully underadvertised, librarians and archivists once again proved to me what a primary role they play in keeping Chicago's architectural and historical legacies not only intact but also remarkably accessible to the public. Chief among these are Michael Featherstone of the Chicago Historical Museum, who led me to some of the city's most important building-stone records, and the Newberry Library's Alison Hinderliter. My thanks are also tendered to archivist Morag Walsh at the Harold Washington Library Center, and to Stephanie Fletcher of the Art Institute's Ryerson & Burnham Libraries.

It was my great pleasure to interact with a number of other persons and organizations dedicated to the preservation of local history in far-flung locations. A morning's conversation and personalized museum tour with Sue Donahue of Illinois's Lemont Area Historical Society taught me much I hadn't known about Chicagoland's own building-stone industry. Out in the Colorado Rockies, Earle Kittleman of the Salida Museum Association conducted at my request extended and fruitful research into the rare but ravishingly beautiful pink granite once quarried in his region. Back in Chicago, Mike Shymanski, president of the Historic Pullman Foundation, updated me on his community's restoration efforts. And I'll never forget the research trips—or pilgrimages, rather—I made to the historic Wisconsin quarry towns of Montello and Berlin. In the former, I met with Kathleen McGwin of the Montello Historic Preservation Society and other local notables, including former quarry owner Bryan Troost. Separately, Bobbie Erdmann, the coordinator of the Berlin Historical Museum, proved to be, in a region of ancient volcanic activity, a volcano herself. But instead of producing the giant clouds of searing ash that had blanketed her region more than a billion and a half years ago, she brought forth a steady flow of reprinted photos, newspaper clippings, and quarry records. And my heartfelt thanks also to Ken Jones, curator of the Amberg Museum, who better informed me of his own community's granite-quarrying history. This Flatlander author is most gratefully indebted to all these paragons of Badger State hospitality and community spirit.

In a separate category lies my debt of gratitude to Michael and Deborah Bill, friends and fellow travelers of many years' standing, who in this particular project provided transportation and other support in its most important phases.

Finally, it's also important to acknowledge the sobering fact that, despite all the crucial contributions cited above, this book is not a definitive survey of Chicago's architectural geology, but merely a first stab at it. No doubt there are mistakes, always mine and mine alone. And there are many other buildings and locales that cry out for attention and recognition. As William B. Lord, writing of his own survey of building stones in an 1887 issue of the *Inland Architect*, noted: "With a field so broad, and information imperfect, and sometimes unreliable, as many facts concerning building stones remain to be solved, my paper can hardly be free from errors and deficiencies, but every effort has been made to avoid them as far as opportunity has permitted."

CHICAGO IN STONE AND CLAY

Introduction

To the person with the heart, soul, and understandings of a geologist, the city is not a denial of nature. It's a vast affirmation of it.

The city is not situated across a yawning ethical gulf from the world of forests, rock outcrops, and pristine prairies. It's a direct and contingent outgrowth of them, still intimately connected to them, and, to a surprising degree, still speaking their language to those willing to hear it. And the city is also civilization's greatest gravitational attractor, drawing to its center an astounding array of local and exotic natural materials, which are then busily assembled by the energies, artistry, and aspirations of thousands of human beings—from robber barons, real estate tycoons, and high-born architects to bricklayers, stonemasons, and mud-spattered caisson-diggers. On one level, it's a showplace of waste, greed, inequity, and horribly squandered power. On the other, it's a well-stocked museum of the Earth's wonders. Chicago, stoked by the short-term success and ultimate tragedy of fossil fuels, may measure its history in hundreds of years; its component rock and clay, in thousands, millions, and even billions.

The grand secret unlocked by both the professional geologist and the amateur rockhound is that the basic materials produced by the Earth's own forces are, in their own way, alive. Like the city itself they are temporary forms that are part of the planet's own dynamic metabolism. The wave-polished stone found on a Lake Michigan beach or the fossil collected in an old Illinois coal-mine pit has a numinous quality that begs for further attention, study, and awe. The humblest common brick, the slab of exhaust-stained granite facing a busy street in the

Loop, also bear epics within them: epics of almost inconceivable units of time; epic accounts of ice age glaciers, vanished seaways, crashing continents, erupting megavolcanoes.

The intrinsic narrative power of clay and stone is amazing. Along Michigan Avenue, for instance, thousands of office workers and tourists stroll by a wall of swirled and spotted rock, never guessing that it's three-quarters the age of our entire solar system. A few blocks farther down, Chicagoans from all walks of life attend concerts and art events in a hall decorated with dazzling polished panels the colors of seafoam and emeralds. This stone began its journey as dense, iron-rich magma miles down, at the crust-mantle boundary. Long after it hardened into rock it was scraped up by a wandering chunk of continent. Chemically altered by contact with saltwater, it was then plastered onto the growing eastern fringe of what is now North America, at a time when the first plants and animals were emerging from the sea. And for every city block beyond these places, more revelations await.

Part I
FUNDAMENTALS

NOTES ON THE BOOK'S FORMAT, AND TIPS FOR EXPLORING CHICAGO'S GEOLOGY

It's been one of the great joys of my life to introduce two generations of Chicagoans to the architectural geology of their hometown. My mission has been to point out a few of the many planes of intersection between the worlds of art, engineering, and science. My conversations and explorations with students and tour-takers have been the inspiration for this book. And these experiences have convinced me more than ever before that this subject, far from being one arcane subset of a single field of science, is in fact an endeavor with the magical property of ever-broadening significance. It opens many unexpected vistas. And it has the unique ability to encourage many different ways of seeing this magnificently incomparable city.

This book is a bridge between the worlds of science and art. It was designed to present a substantial amount of information about both in as engaging a way as I know how. Decades of writing, teaching, and guiding tours have taught me that even the most enthusiastic reader or student will eventually feel overwhelmed by a surging flood of facts, names, and figures about rocks, buildings, civil-engineering stratagems, the lofty flights of architectural genius, natural processes, and the vastness of geologic time. Even this most wonderfully nourishing mixture of subjects can become a deadly brew if presented in too didactic and wholly impersonal a way.

So from the beginning my account was quite consciously designed to have a first-person dimension to it, in which my own experiences and opinions, shared openly with the reader, play a supporting role. I happily admit that it's a hybrid approach to a subject that others, more comfortable writing technical literature for specialists than producing nonfiction prose for a wider audience, have preferred to keep artless, detached, and fully depersonalized. As much as I've profited from these forerunners, I take my cue instead from the naturalists, explorers, architectural critics, and public educators of earlier eras—Alexander von Humboldt, John Ruskin, May Theilgaard Watts, and Donald Culross Peattie come to mind—who knew how to combine the facts at hand with gripping personal narrative or a memorable sense of style. To quote Humboldt himself: "A book on Nature ought to produce an impression like Nature herself. . . . I have endeavored in description to be truthful, distinct, nay even scientifically accurate, without getting into the dry atmosphere of abstract science."

Site Names

The question of whether to cite Chicago buildings by their original or current names is a maddening one. Some sites have collected more monikers over the years than an Austro-Hungarian grand duke. I've concluded the best nomenclatural system is to have no consciously attempted system; to do what seems right in each individual case. However, I suspect that underlying my seeming whimsicality is an instinctive urge to use the name most commonly encountered in the architectural literature. In any case, site descriptions in part II also cite each building's alternative names, if applicable.

The Author Is Not Responsible for Any Missing Buildings or Portions Thereof

The Chicago-architecture historian Carl Condit put it best: "We seem to have arrived at the stage where we consume buildings like clothing and household utensils." Notable sites I featured in tours at the dawn of this millennium are now completely gone or altered beyond redemption, and there's no reason to think that the city's twin mania for building things up and then tearing them down again will cease anytime soon, despite significant progress in developing historic-preservation regulations. I have repeatedly checked all sites described here before publication, but it's a safe bet that not everything described herein will survive, at least with all the features I note, in the passing of the years.

What's Emphasized and What's Not

Many sites described in this book feature more than one geologically derived building material. For each location, I emphasize those materials I have reliably identified over those that are ambiguous or lack sufficient provenance. Sometimes I speculate on the undocumented with ample warning of the speculation; other times for the sake of space restrictions I focus only on the known. Fortunately, there are quite a few sites where multiple, well-identified points of interest can be elaborated upon.

Be Respectful of Private Residential Property

Some of Chicago's greatest geologic treasures can be found adorning houses in many different neighborhoods, from the stratospheric streets of the Gold Coast to far-South-Side Pullman, and from the bustling byways of River North to the leafy confines of Kenwood and Hyde Park. Enjoy these sites, as I have with my tour attendees, but don't jump fences or otherwise trespass to get a better look. In some site descriptions I note that a house's ornamental details or stone types are best seen through some sort of magnification device. But if you do use one, be considerate and discreet. How would you like to look out one of your own windows and see a stranger peering at you through field glasses?

CHICAGO'S GEOLOGIC SETTING AND HISTORY

> The beholder, especially if he be a geologist, feels a strange spell stealing over him. Mighty visions of the old geologic ages enrapture his soul. A leaf from the old stone book is upturned before him, and he reads in the great Bible of Nature her sublime truths. He has discovered hard sense, common sense in the rocks.
>
> —James Shaw, writing in *Economical Geology of Illinois*, 1882

The Geologic Time Scale

Geology is an intensely historical science. Geologists have the same temporal mindset that architectural historians do, but their field of inquiry extends roughly four hundred and fifty thousand times farther back into the mists of the past. If one could make a 1,435-foot linear chart of the history of the Earth and drape it vertically from the 110th and uppermost story of Chicago's Willis Tower all the way to the sidewalk below, the whole span of notable human architecture, from the earliest known Neolithic settlement in Asia Minor to the latest Windy City skyscraper, would occupy less than four hundredths of one inch at the bottom. That's a span too tiny to be marked by even the finest pen. In contrast, the geologic history covered in this one book would extend from the ground more than three-quarters of the way up the entire building.

In demarcating their time line, geologists use the terms denoting its subdivisions quite precisely—much more so than students of the human saga do. While the latter may refer as they wish to the Victorian period or the Victorian era and mean exactly the same thing, Earth scientists have a distinct hierarchy of time that is, so to speak, set in stone. The largest subdivisions are without fail the eons, immense chunks of Earth history in which even the smallest is more than half a billion years in duration. The eons are sliced into eras, which in turn are divvied up into periods, and then, at the lowest level discussed in this book, epochs. Standard abbreviations are used for the huge numbers geologists wield: to

EON	ERA	PERIOD	SUBPERIOD	EPOCH	AGES OF SOME CHICAGO BUILDING MATERIALS
PHANEROZOIC 541 Ma–present	CENOZOIC 66 Ma–present	QUATERNARY 2.6 Ma–present		HOLOCENE, 12 ka–present / PLEISTOCENE, 2.6 Ma–12 ka	In one or both epochs:Tivoli Travertine: Porter and Chicago Brick clays
		TERTIARY 66.2–2.6 Ma		PLIOCENE, 5–2.6 Ma	
				MIOCENE, 23–5 Ma	
				OLIGOCENE, 34–23 Ma	Carrara Marble
				EOCENE, 56–34 Ma	Mokattam Limestone
				PALEOCENE, 66–56 Ma	
	MESOZOIC 252–66 Ma	CRETACEOUS 145–66 Ma			Tinos Ophicalcite
		JURASSIC 201–145 Ma			Alicante, Botticino, Massangis, and Rosso Ammonitico Veronese Limestones; Conway Granite; Larissa and Levanto Ophicalcites; Paltone Breccia; Portland Sandstone; Rapidan Diabase
		TRIASSIC 252–201 Ma			Portoro Limestone
	PALEOZOIC 541–252 Ma	PERMIAN 299–252 Ma			Larvikite Monzonite; Narragansett Pier, Porriño, and Stone Mountain Granites
		CARBONIFEROUS 359–299 Ma	PENNSYLVANIAN 323–299 Ma		Crossville and Dunreath Sandstones; Illinois Basin clays for Chicago Terra-Cotta makers
			MISSISSIPPIAN 359–323 Ma		Buena Vista Siltstone: Burlington-Keokuk, Salem, and Warsaw Limestones: Mount Airy Granodiorite: Napoleon Sandstone; Yule Marble
		DEVONIAN 419–359 Ma			Aberdeenshire, Concord, Deer Isle, Hallowell, North Jay, Quincy, Vinalhaven, and Woodbury Granites; Barre Granodiorite; Berea Sandstone; Euclid Siltstone; Languedoc Limestone
		SILURIAN 444–419 Ma			Artesian, Camp Sag Forest, Lannon, Lemont-Joliet, and Valders Dolostones; Cape Ann and Spruce Head Granites; Stønen Trondhjemite
		ORDOVICIAN 485–444 Ma			Holston Limestone; Oneota Dolostone; Shelburne Marble
		CAMBRIAN 541–485 Ma			Bonneterre Dolostone
PROTEROZOIC 2.5 Ga–541 Ma	NEOPROTEROZOIC 1.0 Ga–541 Ma				Aswan Granite and Aswan Granodiorite; Milford and Westwood Granites; Jacobsville Sandstone (probably); Stony Creek Gneiss (mostly)
	MESOPROTEROZOIC 1.6–1.0 Ga				Graniteville, Town Mountain, and Waupaca Granites; Marcy Metanorthosite; Mellen Gabbro; South Kawishiwi Troctolite
	PALEOPROTEROZOIC 2.5–1.6 Ga				Athelstane, Montello, and St. Cloud Area Granites; Berlin Rhyolite; Impala Gabbro; Lavras Gneiss; Sioux Quartzite
ARCHEAN 4.0–2.5 Ga	NEOARCHEAN 2.8–2.5 Ga				Hinsdale and Milbank Area Granites
	MESOARCHEAN 3.2–2.8 Ga				
	PALEOARCHEAN 3.6–3.2 Ga				Morton Gneiss
	EOARCHEAN 4.0–3.6 Ga				
HADEAN 4.56–4.0 Ga					

FIGURE 2.1. The geologic time scale with representative Chicago building materials listed by age.

them, "Ga" connotes billions of years, or billions of years *ago*; "Ma," millions; and the uncapitalized "ka," thousands. This easily mastered shorthand will be used throughout the book.

As seen in the figure 2.1 time scale, each of these spans of time has been given its own distinctive name. To the Chicago geologist, some are of special significance: the bedrock that rests below us dates from the Silurian period of the Paleozoic era of the Phanerozoic eon, while the glacial sediments above them were deposited in the Pleistocene epoch of the Quaternary period of the same era and eon. And within Chicago's borders lie a multitude of building-stone varieties from many other points on the time line, with the oldest on architectural display being the venerable Paleoarchean-era Morton Gneiss that adorns some of the city's finest Art Deco buildings. In contrast, that workhorse of locally produced building materials, Chicago Common Brick, was made from material from the other end of geologic time—sediments deposited in the latter portion of the Quaternary period.

Origins

The discovery of the age of the Earth, and the elaboration of many of its most important events since, has been one of science's greatest collaborative efforts, involving thousands of men and women working over the course of centuries. Their patient and productive labor has resulted in one of the greatest acts of consciousness-raising humankind has ever known. Various lines of evidence, from the mapping and sequencing of rock units in the field to the study of fossils, glacier ice cores, and ocean sediments have led us to fill in the geologic time line to a significant degree. And from still more exotic evidence, including the isotopic analysis of meteorites, we've inferred that our planet formed approximately 4.56 Ga. In other words, it is, from what cosmologists now tell us, almost precisely a third of the age of the entire universe.

For the vast majority of its history, the Earth wore a much different aspect than it does now. In addition to all the amazing changes over eons in its living communities and climates, it has demonstrated that even its rocky crust is in endless motion in a process geologists call plate tectonics. Over millions of years ocean basins yawn open, then close; continents waltz giddily across the globe's surface and periodically grow through collision with other landmasses or are torn asunder by huge rift zones.

North America had its origin as a much smaller landmass, the Superior Craton, comprising what is now central Canada and most of Minnesota. At various times since the Archean eon other crustal sections were added to it to form the landmass geologists call Laurentia. At 650 Ma, near the end of the Proterozoic eon, Laurentia was situated deep in the Southern Hemisphere, not far from the modern location of Antarctica. In the Pennsylvanian subperiod, some 350 Ma later, our portion of what had become the immense supercontinent Pangaea straddled the equator.

The Paleozoic World and the Formation of the Region's Bedrock

The **Paleozoic era** (541–251 Ma) was the first subdivision of our current, Phanerozoic eon. It witnessed the explosive expansion of multicellular life, as well as a great deal of plate-tectonics activity, and within its span the plant and animal kingdoms diversified dramatically despite periodic mass extinctions. The segment of the Paleozoic of greatest interest to the Chicago geology enthusiast is the **Silurian period** (443–419 Ma). It was then that the dolostone of our region's uppermost bedrock formed from layers of limey mud that had settled to the bottom of a shallow subtropical sea. In this hot and saline environment, perhaps quite similar to the modern Persian Gulf, some of the Earth's first coral reef communities flourished. The remains of these reefs, partially exposed, can still be glimpsed in such places as the great Thornton quarry just south of the city along Interstate 294, and Lemont's Sagawau forest preserve.

The bluish-gray to buff Silurian dolostone extracted from quarries on Chicago's West Side and in the Lower Des Plaines River Valley served as Chicago's first great architectural dimension stone, and it can be seen today in a number of structures built before and fairly soon after the Great Fire of 1871. The most famous examples of these are the Chicago Water Tower and Pumping Station on North Michigan Avenue, where their exteriors, which have weathered to a warm and golden tone, show how attractive a building material this rock type really is.

Brief note should also be made of the **Pennsylvanian subperiod** (323–298 Ma). Pennsylvanian units form the bedrock in much of the rest of Illinois, including areas southwest of the city in portions of Will, Grundy, and Kankakee Counties. There its various sedimentary rock types include both the underclays and shales used by Chicago's terra-cotta and pressed-brick industry and the seams of bituminous coal that, in the period of the city's rapid industrial growth, provided a cheap, abundant, and locally available fossil fuel energy source—albeit one that also produced an immense amount of air pollution, soot, and grime with all the health problems they generated.

The Quaternary Period

If the far-distant Silurian was our city's great early provider of bedrock, it was in the most recent period, the **Quaternary** (2.6 Ma–present), that the thick mantle of unconsolidated sediments above the bedrock was laid down. These sediments range from till, an unsorted mixture of all particle sizes from clay to boulders that was laid down directly below or in front of a glacier during one of the later ice-sheet advances of the Pleistocene epoch (2.6 Ma–12 ka), to lakebed clays and silts deposited when Lake Chicago, Lake Michigan's precursor, rose up to 60 feet higher than its current level, covered much of what is now the city, and created spits (long and gently curving deposits of sand and gravel) that now, as exposed ridges, remain important features of the modern urban landscape. The lake's highstands occurred at the end of the Pleistocene and in early phases of the **Holocene epoch** (12 ka–present).

The Structural Story

When geologists cite the structural branch of their science, they refer to the identification and interpretation of features in the Earth's crust—especially places where the crust has been warped, bent, broken, or moved. While by the standards of the Colorado Rockies or even northern Wisconsin the surface of the Chicago region is subtle, with flat to gently rolling terrain in most locales, our subsurface bears the distinctive imprint of major tectonic forces produced in separate pulses by landmass collisions and mountain building almost a thousand miles to the east, during the Paleozoic era. While the crustal disturbances this far west were by no means as dramatic and complicated as they were in what is now in Vermont's Green Mountains or the Appalachians farther south, the collisions still did generate giant wavelike patterns, with crests and troughs, through the Midwest. In fact, the Chicago region sits atop a relatively narrow zone of strata that have been pushed upward into what is called the Kankakee Arch (see figure 2.2). On either flank are much larger areas where the rock layers are bowed downward instead into the Michigan and Illinois Basins. While these structures are not immediately evident at the surface due to the planing effect of millions of years of erosion, they did play a crucial role in determining the ecological and depositional environments of the past, and hence the economic resources of the present. In the Silurian, for example, Chicagoland's reefs formed on a narrow platform in the shallow water of the developing Kankakee Arch, whereas the sea was considerably deeper in the basins to both the southwest and northeast.

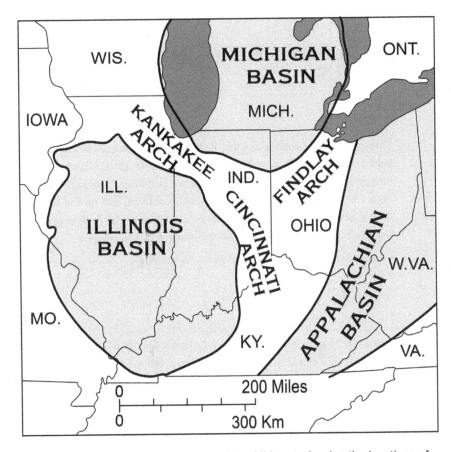

FIGURE 2.2. A basic structural map of the Midwest, showing the locations of two features cited in this book—the upwarped Kankakee Arch and the down-warped Illinois Basin. (© Indiana Geological and Water Survey, Indiana University, Bloomington, Indiana; reproduced by permission)

THE GEOLOGY OF BUILDING MATERIALS

There is no help: sooner or later, in the course of practice, the architect or engineer will have the need of some geological knowledge forced upon him. Force, indeed, is seldom required; for brought as he is into direct contact with geological problems, and having by the nature of his training developed an inquiring mind, the architect or engineer has often become a willing student of the science.

—John Allen Howe, *The Geology of Building Stones*, 1910

Over the centuries human beings have relied upon a host of different materials taken from their local environments to provide themselves with shelter. Some of the earliest of these had a biologic source—from woolly mammoth bones used by Upper Paleolithic hunter-gatherers in Eastern Europe to such plant-based products as woven palm fronds, bundled reeds, and straw for thatch. But wherever substances of geologic origin were accessible, they too were used. Chief among them were clay and sand for adobe, and that better provider of solidity, stone. But as societies became wealthier and more complex, they were no longer tied to local materials. Today, a short stroll in practically any Chicago neighborhood reveals a multitude of stone types from many parts of the globe. But even in ancient times rock types of particular ornamental appeal were imported from distant sources. For example, by the reign of the emperor Augustus (27 BCE–14 CE), ancient Roman architects had at their disposal premium white marble from what is now Carrara, some 200 miles up the Italian peninsula from the imperial capital, as well as yellow African breccia from Numidia, more than twice that distance across the Mediterranean Sea. And in succeeding reigns granites from still more distant Egypt were used in such notable buildings as Vespasian's Temple of Peace and Hadrian's Pantheon.

Building Stone

What's the difference between *stone* and *rock*? Nothing, really, except that geologists tend to rely on the latter term while architects quite exclusively use the former. Building stone can be derived from any rock type deemed suitable for the purpose, and is obtained most usually from quarries, excavations large or small into the Earth's crust where the rock is removed by mechanical means and split or cut into smaller, transportable units.

In some cases, however, stone may also be obtained from loose pieces of rock already freed from their point of origin by the erosive action of glaciers, surf, or flowing streams. For instance, the exterior of the Edgewater neighborhood's striking Epworth United Methodist Church is adorned with glacial-erratic boulders barged down the lake coast from Wisconsin. And if you choose to venture farther afield to Crystal Lake, Rockford, Kenosha, or Port Washington, you'll discover homes of nineteenth-century settlers clad in artfully arranged beach and stream cobbles.

The Geological Classification of Stone

When asked what the science of geology concerns itself with, most nonscientists simply reply, "rocks." Eighteenth-century lexicographer Samuel Johnson preferred to define it instead as "the doctrine of the earth." And indeed the entire planet (not to mention every other rocky or icy world so far discovered) is its proper domain. It's a broad-ranging realm of inquiry with many subdisciplines, including those with well-forged links to biology, physics, climatology, and chemistry. But there's no denying that within it petrology, the study of rocks, still holds pride of place. Petrologists group the great inventory of rock types our world has to offer in three overarching categories: *igneous*, *sedimentary*, and *metamorphic*. The urban naturalist willing to explore Chicago's architecture can readily find outstanding examples of all three.

IGNEOUS ROCKS

With its name derived from *ignis*, the Latin word for fire, it's obvious enough that this group has a fiery or at least hot and molten origin. Whatever its exact chemical composition or appearance, each variety of igneous rock begins with the cooling and solidification of magma (molten rock underground) or lava (molten rock that has reached the Earth's surface before hardening). There are also a few extrusive varieties that form from the deposition of ash spewed into the air by some types of volcanic eruptions. But even here the basic rock-forming material is derived from a recently molten source.

There are two main criteria petrologists use to identify each igneous rock:

- Is it *intrusive* (formed from magma, and with larger mineral crystals, visible to the naked eye) or *extrusive* (formed from lava or ash, with smaller, often microscopic crystals)?

- Is it *felsic* (rich in silicon and aluminum, and lighter in both color and weight), *mafic* (rich in iron and magnesium, and heavier and darker), *intermediate* (with composition, weight, and appearance more or less halfway between felsic and mafic), or *ultramafic* (very rich in iron and magnesium, and very dark and heavy)?

Each of the igneous building-stone varieties to be found on or in Chicago buildings and monuments can be succinctly described and visualized using this system. The most frequently encountered types are:

Granite: intrusive, felsic; this category includes both regular and alkali-feldspar granites

Granodiorite: intrusive, felsic to intermediate

Monzonite: intrusive, intermediate

Tonalite: intrusive, felsic

Trondhjemite: intrusive, felsic

Gabbro: intrusive, mafic

Troctolite: intrusive, mafic

Diabase: intrusive, mafic

Anorthosite: intrusive, felsic

Rhyolite: extrusive, felsic

Tuff: extrusive, usually felsic or intermediate

Using granite as an example, we see that it has individual mineral grains or crystals large enough to distinguish without a magnifier. Also, it's usually lighter in color—frequently pink or a paler shade of gray. In contrast, gabbro is heavier per unit volume, dark trending toward black, but like the granite with grains large enough to see without a hand lens.

Igneous rocks have played a major role in architecture since at least the time of the pharaohs of ancient Egypt. Many types are renowned for their beauty, hardness, durability, and ability to take a high-gloss polish.

SEDIMENTARY ROCKS

This category, widely represented in both Midwestern outcrops and Chicago architecture, is split into two groups, *clastic* and *chemical*. The clastic category pertains to all sedimentary rock that forms by the accumulation and compaction or cementation of sediments that can range from tiny microscopic specks of clay through silt, sand, gravel, and cobbles to hefty boulders. These rocks are most readily distinguished from one another by their particle size. On the other hand, chemical sedimentary rocks form from the precipitation out of water solution of previously dissolved compounds. They're identified by their mineralogical makeup—whether they contain calcite, dolomite, quartz (silica), or other compounds.

The most common sedimentary building stones used in Chicago are:

Dolostone: chemical, composed of dolomite and often some calcite as well
Limestone: chemical, composed of calcite; includes **travertine**
Sandstone: clastic, composed of sand-sized grains; includes **brownstone**
Siltstone: clastic, composed of silt-sized grains
Orthoquartzite: the same as sandstone, but cemented with silica and hence much harder
Breccia: clastic, composed of angular pebbles, cobbles, or boulders; includes
  Ophicalcite: a breccia containing fragments of metamorphic serpentinite and marble

Sedimentary rocks are also notable because they often contain fossils or such other interesting depositional features as ripple marks and crossbedding. Examples of all these can be found in Chicago buildings.

METAMORPHIC ROCKS

This group includes some of the most exotic and flamboyantly patterned stone types and is quite widely used in architecture. As their name implies, these rocks have undergone a metamorphosis of one kind or another—a change wrought when the original rock was subjected to high temperature, great pressure, or hot, chemically reactive fluids circulating underground. For this category geologists use a number of criteria to identify specific varieties, but two are most salient to students of building stone:

- Is the rock *foliated* (with minerals in parallel alignment, producing a banded, wavy, or thinly laminated appearance) or *nonfoliated* (without such parallel patterning)?
- What was the *parent rock* (which could have been igneous, sedimentary, or even another metamorphic rock)?

The varieties of metamorphic building and monumental stone most encountered here are:

Marble: nonfoliated; parent rock either limestone or dolostone
Serpentinite: nonfoliated to weakly foliated; parent rock dunite or other forms of peridotite
Soapstone: nonfoliated; parent rocks include peridotite and serpentinite
Gneiss: foliated; parent rock granite or other igneous, sedimentary, or metamorphic rocks
Slate: foliated; parent rock shale
Metanorthosite: weakly foliated; parent rock anorthosite

Migmatite: a foliated "mixed rock" containing both gneiss or another meta-morphosed rock and younger, unmetamorphosed granite or tonalite

Metamorphic rocks range from the tough, hard, and enduring (gneisses and slates) to those that, for all their appeal, are made of minerals much more susceptible to weathering in the Windy City's brutal climate (marbles, serpentinites). The latter are best used to adorn interior spaces—which is not to say that some daring or geologically innocent Chicago architects haven't employed them, sometimes quite extensively, outdoors.

The Architectural Classification of Stone

We have frequently pointed out in these columns that there is a regrettable looseness and inaccuracy in the stone trade in the classification of the different varieties of stone. Some of this is due to pure ignorance, doubtless, but a greater part of it arises from a commercial spirit, a desire to call to the aid of merchandizing a more sounding and alluring name, or an attempt to differentiate a particular stone from those of its own class.

—from a 1912 editorial in the trade journal *Stone*

The student of architectural geology soon comes across a major stumbling block of terminology: architects and contractors simply do not use the carefully elaborated geological system of rock names outlined above. However bothersome this may seem to a scientist, it must be remembered that architects and builders travel along much different educational and professional arcs that for all of their rigor and demands include no training in geology—just as most geologists couldn't tell you the difference between a plinth and a pepperoni pizza.

At first glance the architectural classification of stone seems much simpler, but it soon reveals itself to be both frustratingly vague and inconsistent. As a general rule the building-trades moniker "marble" refers to *any* softer rock type that can take a high polish. "Granite" denotes *any* harder rock that also takes a good shine, and it can refer not just to igneous types but metamorphic and sometimes even sedimentary ones, too. And the term "stone," to the geologist hopelessly generic, refers to those remaining rock types that can't be polished. Worse yet, various exceptions to all these definitions are soon encountered.

Sometimes preservationists and architectural historians also try their hand at the more detailed terminology of the geologic system. But in doing so they often misidentify the stone in question or create false synonyms—for example, making sandstone and limestone interchangeable terms for the same rock. One can find various examples of this sort of semantic sloppiness in even the most authoritative architectural books and historic-preservation reports.

For the sake of both consistency and accuracy, this book uses a stone's official stratigraphic or unit name when one has been assigned and is still accurate. If an official name is lacking, I usually identify the stone by its place of origin. In the case of foreign stone selections that are not identified by published designations, I adopt the most reasonable (or, at any rate, the least hyperbolic) version of their trade names, especially if they refer to the stone's locale of origin. These preferred names are capitalized, as in Morton Gneiss and Carrara Marble. And in every instance I've tried my best to ensure the stone's rock type is correctly identified. In the text I often also cite the stone's most common trade names, capitalized and set in quotation marks. Accordingly, when discussing Chicago's most common rock type, the Salem Limestone, I often note it's widely known in the building trades as "Bedford Stone" and "Indiana Limestone."

One special situation has prompted me to diverge a bit from the approach outlined above. For Chicagoland's own bedrock, the dolostone variously quarried from the Racine, Sugar Run, and Kankakee Formations in Illinois and corresponding units in Wisconsin, I've chosen to give the reader a better sense of where exactly each site's stone was produced. I do so by parsing, whenever possible, my general **Regional Silurian Dolostone** category into several more specific varieties:

Lemont-Joliet Dolostone, from the Lower Des Plaines River Valley of northeastern Illinois

Artesian Dolostone, from Chicago's West Side

Camp Sag Forest Dolostone, quarried by Civilian Conservation Corps crews in the 1930s on the grounds of what is now Lemont's Sagawau Environmental Learning Center

Lannon Dolostone, from Lannon, Wisconsin

Valders Dolostone, from Valders, Wisconsin

While it can be difficult to quarry and expensive to transport over large distances, stone has been a preferred building material for millennia. Its ability to impart a sense of solidity, permanence, and even grandeur to an architectural design is unparalleled. And it's often remarkably beautiful in its own right. For untold generations, architects have carefully studied and artfully employed the ornamental attributes of marbles, granites, and many other rock types.

The process of finishing, or transforming a piece of rock extracted from the headwall of its quarry to a fully prepared unit of building stone, involves a number of steps, tools, and methods to produce the desired effect. First, the quarried block is split, sawn, or otherwise shaped into the preferred dimensions. Abrasives

FIGURE 3.1. Silurian Lemont-Joliet Dolostone outcropping in the Lower Des Plaines River Valley, where it was previously quarried as Chicago's first widely used building stone.

may then be applied to its outward-facing side to produce either a nonreflective honed or reflective polished surface. Also, a pleasing ornamental effect may be achieved with sandblasting or the application of a water jet, an acid wash, or a high-temperature flame. Or the surface can be scored with metal brushes, hammering tools, or chisels. All these techniques give the architect a wide assortment of effects and textures to choose from. Some of these have proven to be quite indicative of a particular style or era. For instance, when the outer surface of a stone block is intentionally chiseled only on its margin and the central portion is left rough, jagged, and projecting, the stone is called rock- or quarry-faced. Buildings sporting this massive, brooding, swords-and-sorcerers look were especially in favor during the Richardsonian Romanesque period of the last decades of the nineteenth century.

While stone has long been preferred for its generally enduring nature, its different types, with their widely varying chemical compositions, undergo the destructive process of weathering at dramatically different rates. The perfect place to observe differential weathering is in Uptown's famous Graceland Cemetery. There marble headstones dating from the mid–nineteenth century have deteriorated so much in the city's extreme climate and polluted atmosphere that quite a few now resemble dissolving lumps of sugar and are quite illegible. A few feet away stand contemporaneous monuments of polished and still pristine granite that look as though they were installed the day before yesterday. As this demonstrates, true marble, for all its ornamental appeal, can be an unwise or even disastrous choice when exposed to the elements. The naturally occurring carbonic acid found in even the most unpolluted rainwater reacts with the calcium carbonate in the marble and can degrade it significantly. However, that doesn't mean this sought-after architectural stone can't be used to stunning and long-lasting effect indoors.

Almost always, orthoquartzite is the building stone that is the toughest and most resistant to weathering, but it can't take a polish and its extreme hardness usually makes it very hard to cut, shape, and work. Following fairly close behind it are the true granites, which are much more amenable to finishing and which last indefinitely, especially when sealed with a good polish or flaming. In contrast, such commonly used sedimentary rocks as sandstone and limestone can be more problematic. The once widely popular Portland and Lake Superior Brownstones tend to spall and chip away easily, especially when not mounted properly or seasoned sufficiently before use. By the same token, limestone's primary chemical constituent is calcium carbonate, which can lead to the same problems cited above for its metamorphic equivalent, marble. That said, one limestone has long been touted for its admirable survival skills: Indiana's renowned and architecturally ubiquitous Salem Limestone. Still, while it shows remarkable resistance to air pollution and most other forces that affect other stone types, it too has its vulnerabilities. When it is used at or near sidewalk grade it wicks up water and dissolved deicing salts. This leads either to exfoliation, where sections of the stone peel away, or to efflorescence, where a white crust accumulates on the surface of the stone. Exfoliation has sometimes also been a serious problem with Chicagoland's own Regional Silurian Dolostone.

Brick

Even in places where locally quarried building stone is abundant, brick has been a successful competitor. From a technical standpoint, bricks are mass-produced, stackable units of dried or fired clay, sometimes mixed with other materials. In their own way they're as geologically derived as rock products are. The earliest recorded use of sundried brick (adobe) dates from the interval 10,000–8,000 BCE; and the oldest evidence of the much more enduring fired brick comes from 5,000 to 3,500 BCE. The ancient Mesopotamians developed the art of molding and glazing ornamental bricks, and in Rome, the architectural use of brick began, somewhat sporadically, in the reign of Augustus, though these early bricks tended to be broken clay tiles sawn into shape. Later, however, imperial architects developed a number of standard sizes of mass-produced fired bricks and used them in various applications, including arches and facing for their concrete structures.

By the nineteenth century, most bricks produced in northeastern Illinois were made directly from clay dug from the region's riverbanks, thick deposits of glacial till, or ancient proglacial lakebed deposits. In other parts of the Midwest, the soft and clay-rich rock known as shale has also been used. The clay, whatever its source, was either molded by hand in wooden forms or run through pressing machines, then dried and placed in kilns for firing. Some of America's greatest cities—Philadelphia, Baltimore, Chicago, Milwaukee, and St. Louis—became major production and distribution centers of brick, but even small farm towns often had their own claypits and brickmaking concerns, often as first-generation industries begun soon after initial settlement.

Brick can be named and identified in a number of ways. One of the simplest is by its quality: the higher grade being *facing brick* and the lower *common brick*. The former term refers to the finest and most attractive types suitable for façades and other places where their ornamental assets can be shown off; the latter, to those that were less carefully made, and hence softer, more pebbly, or more porous. Common brick was often made from glacial till, whereas facing varieties were fabricated from more well-sorted alluvial (stream) deposits or Pennsylvanian-subperiod shales and underclays (ancient soils, also known as paleosols). Traditionally, they've been relegated to more utilitarian and less visible uses, to side or back elevations, or basement and other interior walls.

Another obvious way of identifying bricks is by their size, shape, or texture. While there are too many variations on this theme to mention here, three specific types are worthy of note because they're cited in several of this book's site descriptions:

Roman Brick is a facing variety that owes its name to the fact it mimics, to some extent at least, the shape of a long and thin brick type used by ancient Roman architects. Some of our region's most eminent designers, including Louis Sullivan, Frank Lloyd Wright, and George Elmslie, have used this variety to stunning effect.

Molded Brick enjoys an even more ancient pedigree, having been first extensively used, and perhaps invented, by the Babylonians some 3,000 years ago. The term refers to any brick, glazed or not, that is custom-shaped to present on its outward face heightened relief or a curved or slanted surface. When set together in a wall, the units of Molded Brick produce an image, pattern, or other ornamental effect.

Rock-Faced Brick has rough, projecting outer surfaces that simulated rock-faced stone ashlar.

Some brick types have coatings that, besides providing a pleasing ornamental effect or a protective layer, can also be used as identification traits. Glazed brick has had a its outer coating fired onto the clay at a relatively lower temperature and is liable to eventually develop chips or the network of fine cracks known as crazing. But in enameled brick the firing temperature is so high that the coating substance directly fuses with the clay to produce a much more enduring seal.

For the architectural geologist, the discovery of a beautiful and exotic building stone in some elegant hotel lobby or office-building breezeway can ultimately prove to be a frustrating experience if it can't be reliably identified. But trying to conclusively name more than a few of the multitude of brick types found in Chicago can be even more perplexing. The majority of them simply can't be sourced, for the same reasons a mystery stone can't. Either there are no surviving construction records or the brick lacks unequivocal identification traits. Fortunately, there is across the city a smattering of sites—and in the case of Pullman, one entire district—where the brick can be named with a high degree of confidence or even certitude.

These are the most identifiable types of brick to be found within our city's limits:

Chicago Common (from Chicago-region yards; soft common brick; buff to peach and salmon, and even ochre or reddish; often contains dolostone pebbles or bits of gravel)

Chicago Anderson Pressed (high-quality facing brick of various colors produced by the Chicago Anderson Pressed Brick Company)

Tiffany (not to be confused with Tiffany Glass; from Momence, Illinois; both high-quality enameled brick and regular pressed brick of various colors)

Cream City (from Milwaukee, Wisconsin, or neighboring towns; soft common or harder facing brick; sometimes described as white but characteristically cream to light yellow; the common variety often contains pebbles or gravel)

Lake Calumet (from the bed of Lake Calumet on Chicago's South Side; relatively soft common brick; buff to ochre or even dark brown)

Porter (from Porter, Indiana; hard facing brick; most often a striking orange red to brick red)

St. Louis (from St. Louis, Missouri; hard facing brick; usually red but the shading varies from pink through bright cherry or brick red to brown and maroon with iron flecking)

Logan (from Logan, Ohio; hard facing brick of a buff or grayish-brown color)

Winslow Junction (from Winslow Junction, New Jersey; cream or pale yellow)

Brick is such a common building material that it invites our boredom. We forget that in the settlement of new American towns the spade that excavated a community's first clay pit was rarely far behind the axe and the plow. Early brickyards in Chicago and elsewhere provided architects with what was essentially convenient, preshaped units of metamorphic rock made of naturally occurring sediments artificially lithified by the intense heat of the kiln, which can reach more than 2,000° Fahrenheit.

Besides being relatively inexpensive and easily transportable, brick is much more fireproof than most types of stone. And it possesses outstanding ornamental attributes when architects and bricklayers conspire to make it do so. As well as choosing types of brick that have special ornamental properties or that alternate with those of a different color, shape, or texture, they can create amazingly elaborate patterns by varying how bricks are set in their courses (rows). These can be composed of those laid horizontally as stretchers or rowlock stretchers, vertically as sailors or soldiers, on end as headers or rowlocks—or any combination thereof. The wide variety of designs that can be so produced seems almost endless, especially when molded or glazed bricks are added to produce intricate ornament and vibrancy of color. One of the better-known ornamental patterns is the Flemish Bond, where each course is composed of alternating stretchers and headers.

Terra-Cotta

Chicago is one of the world's great showplaces for terra-cotta. Literally translated from the Italian as "cooked earth," its primary ingredient is the same as brick's—clay. In modern times we find it in everything from flower pots to roofing tiles, and, in its most impressive form, as glazed and custom-molded units of building cladding and ornament.

One subset of the terra-cotta category deserves particular mention. Ceramic tile, familiar to anyone who has ever been in a bathroom, consists of relatively small, thin squares or rectangles of glazed clay, from glistening pure white to polychromatically patterned, used for decorative cladding. Employed most gloriously in Islamic North Africa and the Middle East, it can nevertheless be found as a major ornamental element in Chicago, too.

One of the reasons so many of the gems of Chicago architecture built in the late nineteenth and first half of the twentieth century feature this building material so prominently is that while Italy has been the world center of terra-cotta artistry and production since medieval times, northeastern Illinois became one of its worthy competitors for several decades starting in the 1870s. Among the firms producing it here were the Chicago Terra Cotta Company, the city's first; the Northwestern Terra Cotta Company, located on the east bank of the Chicago River at Clybourn and Wrightwood Avenues; the American Terra Cotta Company, situated in Terra Cotta, Illinois, now part of Crystal Lake; and Midland Terra Cotta Company, of Chicago. And other firms, local or not, specialized in terra-cotta roof tile. These included the Ludowici Terra Cotta Roof Tile Company, of Chicago and later other locations as well; the Celadon Terra Cotta Company, of Alfred, New York and later other locations as well (eventually merged with Ludowici); and the Akron Roofing Tile Company, of Akron, Ohio. And still others produced ornamental ceramic tile, also known as faience: the Rookwood Pottery Company, of Cincinnati, Ohio; and the Grueby Faience Company, of Revere, Massachusetts. These producers hired hundreds of workers and superbly skilled artisans, many of whom were recent immigrants from Italy and other parts of Europe. They brought with them a centuries-old artistic legacy.

The classification of terra-cotta can be done by its basic properties (unglazed or glazed), or by its type of use (roofing tiles, cladding, or faience). Also, much like brick, it can be identified by the firm that made it, if known. In this book, terra-cotta produced by the companies mentioned just above is indicated wherever the historical record permits.

At a time when Chicago was rebuilding and expanding after the Great 1871 Fire, the virtues of terra-cotta certainly did not go unnoticed by architects. For one thing, its relatively light weight made it less expensive and less cumbersome to transport than building stone. For another, its resistance to high temperatures and flame made it easier for builders to comply with new fireproofing regulations, imposed by insurance companies and city codes alike, by sheathing their iron or steel structural elements in it. And from an aesthetic standpoint, it was hardly a poor second to any other building material. Not only could it be cast into remarkably intricate patterns, as the miraculously ornate terra-cotta moldings Louis Sullivan designed testify; it could actually mimic choice building stone so closely as to be practically indistinguishable from it. One must carefully scrutinize a gleaming white wall in a Loop office-building interior to tell whether it's really choicest Carrara Marble or its ceramic simulacrum. Even granite, with its mosaic of multicolored crystal grains, came to be copied with amazing precision by the use of an ingenious "Pulsichrome" technique of terra-cotta patterning. Even a professional geologist is likely to be hoodwinked by this would-be igneous rock unless there's a telltale crack or chipped panel nearby that exposes the bisque, the yellowish baked clay under only a fraction of a millimeter of glaze.

Concrete

This everyday substance is actually a wonder concoction with fascinating chemical properties. In modern times it's most often a blend of lime- and clay-containing Portland Cement and aggregate that, when mixed with water, sets into stonelike form in an exothermic (heat-releasing) reaction. Concrete is so ubiquitous in modern times it could be fairly said to shape our built environment to an extent no other substance does. We walk on it, we fill caissons and erect walls with it, and even prefabricate and modularize it. The ancient Romans mastered its applications for everything from aqueducts to bridges, and, as their empire endured, increasingly depended on it for their most ambitious civic structures, including the awe-inspiring, concrete-domed Pantheon. They employed a recipe considerably different than our own—one indispensable ingredient being the volcanic ash called pozzolana. But in fact concrete had already been in use for several millennia before the Romans perfected their version of it.

In this book, concrete is not finely parsed into different categories, but special mention is made when it has been given an unusual surface texture.

Since at least 1874, with the construction of the Loop's Delaware Building, architects have not hesitated to use concrete for building façades as well as their foundations. Easily molded, generally durable, structurally strong, and relatively inexpensive, concrete can also be remarkably attractive.

Plaster-Based Surfaces

Plaster is another building material with both an ancient origin and a certified geologic pedigree. While its composition is variable, it always incorporates a chemical compound serving as its binder. Traditionally, this was either the mineral gypsum (calcium sulfate) or lime (calcium oxide and hydroxide) derived from the mineral calcite, otherwise known as calcium carbonate. Lime-containing cement, including Portland Cement, has also been used as a binder. In addition, plaster requires water and sand or some other aggregate. Some aggregates have the special property of either quickening or delaying the hardening process.

When utilized to coat or ornament a building's exterior, plaster is called stucco; when used to bind brick, stone, or units of terra-cotta together, it's mortar. And when it's blended and finished to simulate polished marble or other fancy stone, it is termed scagliola (pronounced scal-YO-lah). While stucco is usually found on houses and other smaller buildings away from the city's center, scagliola takes its place along with stone and terra-cotta in some of the city's greatest architectural landmarks. And decorative mortar transcends the humble role of simply being a binder for other materials by producing a color contrast or other pleasing visual effect.

Metal Materials

The metals used on Chicago's buildings also have undeniable geologic origins and first formed in the Earth's crust in various ways. In modern times, iron is most frequently mined from ancient Proterozoic or Archean sedimentary rocks, such as those found in northern Minnesota and Michigan's Upper Peninsula. Other metal ores accumulated in sedimentary stream deposits or were emplaced by mineral-rich, hot-water solutions infiltrating preexisting rock units. These metals have been used on roofs, building façades, external and internal ornament, and sculpture:

Copper
Bronze (an alloy of copper and tin or arsenic, including the variant **Gilt Bronze**, in which a thin layer of gold is applied to the surface)
Brass (an alloy of copper and zinc; including the variant **Naval Brass**, an alloy of copper, tin, and zinc)
Beryllium-Copper Alloy
Cast Iron
Nickel Silver (an alloy of copper, nickel, and zinc that can resemble real silver or be pale yellow instead)
Stainless Steel
Weathering Steel
Aluminum (including the variant **Anodized Aluminum**, treated to produce an attractive black or colored surface on the metal)
Pressed Metal (sheeting of any suitably malleable metal or alloy, stamped with decorative designs)

One other metal building material that has played a huge role in Chicago's architecture and civil engineering is normally not on view to the urban explorer. Nestled within its sheath of concrete, rebar (a contraction of reinforcing bar) is generally seen only during a building's construction phase, or years later, when it has corroded, expanded, and become exposed as the cracked concrete around it spalls away. Most usually the rebar is made of rust-prone carbon steel, though more enduring and considerably more expensive types are also used in some projects. The purpose of using rebar regardless of its composition is to compensate for concrete's Achilles' heel. While it has impressive compressive strength (resistance to crushing), its tensile strength (resistance to being pulled, bent, or twisted) is considerably less when not reinforced.

Ornamental Glass

Glass, too, is a geologically derived material that features three different types of compounds: a former, silica (the mineral quartz, usually obtained from sand); fluxes such as potash (potassium carbonate) or soda (sodium carbonate); and a stabilizer, most usually calcium carbonate derived from limestone. Glass can also contain as coloring agents either metallic oxides or other substances.

While this book makes no attempt to point out the scientific significance of the untold myriads of window panes on the buildings it describes, there are two special types of glass, Venetian and Favrile, that are worthy of note because of their use in the city's mosaics and other architectural decoration. The first of these is a product of Murano, an island in the Lagoon of Venice archipelago of northeastern Italy renowned for many centuries as a producer of fine glass. Favrile Glass, on the other hand, was invented in America by Louis Comfort Tiffany. It features a striking play of colors and an opalescent luster.

FOUNDATIONS

Engineering in a Birthplace (and Worst Place)
for Skyscrapers

A Deep Look under Chicago's Streets

The Roman author Marcus Vitruvius Pollio stated in his famous treatise *On Architecture* that there are three cardinal virtues for any building: *utilitas* (usefulness), *firmitas* (soundness and stability), and *venustas* (an appearance pleasing to the eye). Here, in what many claim is the hallowed birthplace of the skyscraper, the most consistent challenge has always been *firmitas*. Blame it on the Windy City's geology.

Every day of the year countless Chicagoans traverse the city's streets and sidewalks unmindful of what lies even a foot or two below the pavement. But both geologists and civil engineers are taught to visualize the landscape as a three-dimensional construct: depending on the profession, either to better understand the record of the past, or to assess the properties of the substrate into which the foundations of very large and very heavy buildings will be laid.

It's thoroughly ironic that Chicago served as the home of many of the earliest skyscrapers because its bedrock—generally the best thing to anchor any hefty building's foundations in—is hidden deplorably deep underground. In the monumental heart of New York City, skyscraper construction has been largely constrained to two separate locales, the Wall Street district and Midtown, places where solid metamorphic rock, the Manhattan Schist, crops out or is at least much nearer the surface. But in Chicago's focal points of high-rise construction, the Loop and the Magnificent Mile, the bedrock lies under a blanket of often waterlogged and oozy sediments that has been likened to the layers of a soggy jelly cake. This extends downward from about 55 to over 100 feet. Early in the city's history architects and engineers learned, sometimes the hard way, that the foundations of the city's larger buildings required clever and effective solutions to avoid embarrassing sags and tilts of walls, floors, and stairwells.

One instance of how disastrous a poorly thought-out stabilization scheme could be was the city's second federal building, known as the US Custom House, Court House, and Post Office. Completed in 1880, it was set on a continuous shallow concrete foundation that before long settled unevenly and cracked, with various problems, from peeling plaster to plumbing breaks, developing above-ground. By the 1890s the massive structure was considered overtly unsafe and was subsequently demolished. As the 1893 Rand, McNally guidebook to Chicago picturesquely put it,

> The dark Gothic mass which arose was too heavy for the soil, and sank steadily. It was bolted together, but continued to sink, amid the lamentations of office-holders who would not flee from the fate they feared. Courts have adjourned precipitously on the loud report of an opening to the walls, or the flooding of a water-pipe, and the tiles in the floors respond with a melancholy rattle as the citizen hurries through the corridors to escape the Post Office draughts.

Ironically, much of that ill-fated structure's dimension and ornamental stone was salvaged for use in the construction of Milwaukee's magnificent St. Josaphat Basilica, where it can be seen today.

Seven decades later, at the site earmarked for the mighty John Hancock Center, the inattentive pouring of concrete and faulty caisson-lining installation resulted in voids and unwanted sediments in the underlying caissons that might have later caused a major collapse of the 100-story skyscraper and untold devastation along North Michigan Avenue. Fortunately, an alert worker discovered telltale signs of the problem before the building went up. At the expenditure of a great deal of extra time and money, the issue was corrected, and for the past half century Big John has stood majestically on its secure footing.

The exact composition and thicknesses of the layers of till and lakebed sediments varies from neighborhood to neighborhood and even block to block, as does the depth to bedrock. But generally, as one descends downward from street level, there is a hard but thin clay crust a few feet deep that some sources incorrectly call the hardpan. Below it is soft, watery clay, then a layer of somewhat firmer clay at about 15 feet down, then still more layers of softer sediment until one reaches the true hardpan, the zone of clay 55 or more feet down that is compressed into much greater stiffness and much less water content. As we'll see, the hardpan has often served as a sufficiently reliable terminus for caissons and deep pile foundations. But for excavators wishing to continue down to the bedrock itself, there is a formidable obstacle. Between hardpan and dolostone lies a stratum of sand, silt, and boulders so saturated with water that it has sometimes been called quicksand.

A Brief History of Chicago Foundations

In his *Four Books of Architecture*, the great Italian successor to Vitruvius, Andrea Palladio, observed that a foundation "sustains the whole edifice above; and therefore of all the errors that can be committed in building, those made in the foundation are most pernicious, because they at once occasion the ruin of the whole fabric, nor can they be rectified without the utmost difficulty." And "the architect should apply his utmost diligence in this point; inasmuch as in some places there are natural foundations, and in other places art is required."

If there has ever been a place where the art of savvy foundation design is an absolute necessity, it's Chicago. But that art was acquired gradually and in stages, as civil engineers and architects grappled with the challenges of erecting ever larger and heavier structures in the city's treacherous substrate.

The earliest types of underground support for the larger buildings were shallow foundations, set in the relatively hard, crustlike surface layers. They came in a number of variants. One featured a solid raft of concrete that extended across most or all of the building's base. Another, employing isolated spread footings, relied on a network of much smaller, individual pads spaced to distribute the structure's weight as equitably as possible. This latter method, advocated by the influential Chicago architect Frederick Baumann (1826–1921), represented the first innovative attempt at solving the deadly serious problem of providing stability in a queasy medium. Originally the individual pads supported pyramids of dimension stone and rubble from which a building's structural piers rose. Unfortunately, they were broad and bulky, and usurped much of a structure's basement volume—space that increasingly was needed for boilers and machinery. This problem inspired flatter, lighter, and space-saving grillage pads consisting of crisscrossing sets of horizontally laid rails or I-beams embedded in concrete. Some other methods were explored, as well.

The dramatic example of the Custom House cited above illustrates the dangers of relying on a whole-area raft foundation. The substrate is all too likely to settle differentially below it, and the structure above might subject different parts of the raft to significantly different weights. But even when separate pyramidal or grillage pads and piers were used, settling—often quite uneven—was almost inevitable. Architects compensated by setting building entrances considerably above grade so they would not eventually sink beneath it. But even the most carefully engineered edifice could end up with tipsy staircases and walls and floors out of plumb. Clearly, a more secure form of building anchoring was needed, especially as both the height and mass of new Windy City building designs rose dramatically. Hence the development of deep foundations in the forms of either very long piles or caissons.

Driving long piles to hardpan or even bedrock has often been a very effective deep-anchoring method, and, according to a 1948 source, it has been most frequently used in the high-rise district north of the Chicago River. The traditional material of choice used was the 50-foot timber pole, but in the 1950s one of the Loop's architectural masterpieces, the Inland Steel Building, became one of the first to rest, appropriately enough, on steel piles driven down 85 feet into direct contact with the Silurian dolostone.

As effective and straightforward as this method seems, it has never been without its problems. The act of driving timber or steel into the substrate creates considerable noise and vibrations that are at best annoying to those in the neighborhood and at worst destructive to the foundations and walls of buildings nearby. It was this potentially serious, litigation-prone drawback that prompted the caisson method to be developed as a less percussive alternative, especially for construction sites with extant structures close at hand. But there is also one much more recently developed method, at least technically considered a form of pile stabilization, that can also be less bothersome to neighbors. Known as a drilled shaft foundation, it uses a "CFA"—a continuous flight auger. This high-tech mechanical excavator digs a deep and narrow well into which concrete is injected as the auger is retracted.

In contrast to the usual pile-driving technique, the process of laying Chicago's famous caisson foundations is minimally invasive. This technique had long been used in American bridge construction, and apparently its first application to a city office building was in 1890 for City Hall in Kansas City, Missouri. But the man who developed this technique in Chicago was William Sooy Smith (1830–1916). Smith was a West Point graduate who, after an early stint as a civil engineer, went on to serve in the federal army as a brigadier general during the Civil War. Later, he tried his hand as a gentleman farmer in the Oak Park area, and then decided to reactivate his prewar engineering career. In doing so he became one of America's most sought-after designers of foundations and bridges. His Windy City projects were many; they included the Auditorium Building and what is now the Chicago Cultural Center. But he first employed the game-changing caisson method to anchor just one portion of the famous 1894 Chicago Stock Exchange designed by Adler and Sullivan. Piles were used for the rest of that now tragically demolished building.

As employed in this city, caissons are deep and watertight shafts, traditionally 6 or 7 feet in diameter, dug to the stiff, deep layer of till and thus dubbed hardpan caissons, or to bedrock, as rock caissons. Lined with hardwood lagging secured with iron rings that was removed only when the concrete was finally poured in, the shafts were also often flared out at the bottom into a conical "bell," especially if they did not reach down all the way to the bedrock. This widened terminus provided an especially solid footing. Until the middle of the twentieth century all the many caissons needed at each site were laboriously excavated by hand using three-man crews consisting of the supervising signal man or headman, the dumper, and the hand miner in the shaft itself. The crew's progress downward was measured in sets—units of depth corresponding to the height of the slats used to line the caisson's wall. Normally the slats were 5 feet 4 inches long, but shorter ones of as little as 3 feet 6 inches were used if the substrate was especially soggy. Each caisson team was reportedly expected to dig three sets per day for a total of 16 feet, but George Manierre, writing in 1916, noted that in an eight-hour shift 11 feet of progress was typical in softer clay, and only 5 feet in hardpan. The quality of concrete poured into the caisson shafts was of utmost importance, as the Hancock Center's near-fiasco mentioned above indicates. To quote Manierre, "Recently buildings have been known to sink on caissons, where on examination it was found that the contractor had not put in the proper quantity of cement." He cited the correct mixture: one part cement to two parts coarse-grained "torpedo" sand and four parts crushed stone.

The hand digging of caissons was still in use when the foundation was laid for the landmark Prudential Building completed in 1955. But the Loop had its first example of machine-drilled caissons just a few years later, when the Borg-Warner Building was erected along South Michigan Avenue. This new technique also introduced steel casings as a replacement for the traditional timber lagging. By this time shafts were routinely continued all the way to bedrock. And now in some cases even the uppermost section of dolostone itself is cored out, to create rock-socket caissons that provide an even more secure connection.

Part II

EXPLORING CHICAGO'S NEIGHBORHOODS

THE LOOP

Northeastern Quadrant

5.1 Michigan Avenue Bridge Houses

Intersection of N. Michigan Avenue and Wacker Drive
Completed in 1920
Architects: Edward H. Bennett; Thomas G. Pihlfeldt and Hugh Young,
  engineers
Geologic features: Anthropogenic stream reversal, stream channeliza-
  tion, Salem Limestone

This is the locale where the City of Chicago began. But both the river and the land adjoining bear no resemblance to what the occupants of the Fort Dearborn stockade, situated just across what is now Wacker Drive, beheld. The surface has been graded up to a commanding height and the river channelized into a great concrete-ribbed trench with its water moving opposite to nature's intended direction, thanks to the opening of the Chicago Sanitary and Ship Canal in 1900. In the city's earliest days, when the fort still stood nearby, the stream channel took a hard turn to the right just a little east of what is now Michigan Avenue, and flowed southward parallel to a deposit of sand formed by the lake's longshore drift and current. The channel continued down almost half a mile to where at last there was an outlet into the open water. But now the river has been bullied, for at least one moment of geologic time, into a much straighter course.

On most days the river crossing at North Michigan and Wacker is a great pedestrian bottleneck, with shoppers and tourists jostling like flocks of starlings along the

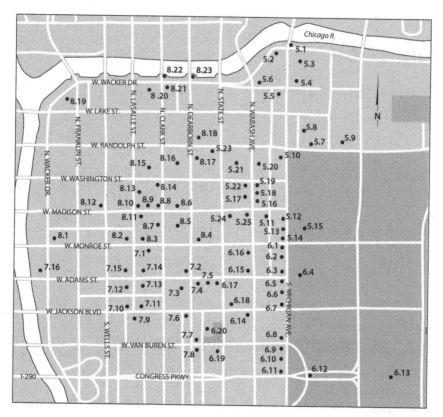

MAP 5.1. The Chicago Loop.

narrow walkways that connect the Loop with the Magnificent Mile. Now officially dubbed the DuSable Bridge, the span here is of the double-leaf (two-sectioned) trunnion bascule design. Many early Chicago River bridges were, in contrast, of the swing type, pivoting on a stream-cluttering center support from perpendicular to parallel with the channel whenever a high-rigged vessel needed to pass. But this more efficient design has the halves of the roadway part in its middle and tilt, like two facing office workers leaning far back in their swivel chairs. It's a beautiful sight when one or both of the sides rise with seemingly effortless grace into the air. The crowds waiting to cross pile up on both sides, watching, a little impatiently perhaps, a tall-masted yacht swan its way through the gap.

At this vantage point of the bridge's southern end the eye is surrounded by a superabundant inventory of stone types few mountain passes or rocky-desert vistas could rival. We might as well start with the commonest of the lot. On the span stand four elegant, Beaux Arts bridge houses. Even on close examination their sandpapery-textured, drab-gray façades suggest they're made of fine-grained concrete. But this in fact is natural rock—and indeed, the one

FIGURE 5.1. A busy day for the Michigan Avenue Bridge. Pedestrians watch a passing sailboat; the upper portion of its mast is just visible under the raised southern leaf. The elegant Beaux Arts bridge houses here feature Salem Limestone, the most frequently used rock type in American architecture.

most widely used dimension stone in the United States. It's the famous Salem Limestone, which in the building trades has been long known instead as "Bedford Stone," though Hoosier quarrymen, proud of the astounding ubiquity of this product of their state, prefer to market it simply as "Indiana Limestone." Technically, this easily overlooked but eminently reliable sedimentary rock is a biocalcarenite, the type of limestone composed of myriad bits of ancient life—everything from shell fragments to tiny whole organisms—cemented together with the mineral calcite. Few persons here ever stop to examine it, but those who do can quickly spot the remains of invertebrate sea creatures who thrived in the shallow carbonate shoals, tidal inlets, and coastal lagoons that were connected to the warm subtropical sea covering southern Indiana, and indeed much of the Midwest, in the Mississippian subperiod of the Carboniferous.

More obvious at first glance on the two bridge houses closest at hand are Henry Hering's sculpted reliefs "Defence" (on the western structure) and "Regeneration," respectively depicting the Fort Dearborn Massacre and the reconstruction of the city after its 1871 Great Fire. Both demonstrate how amenable the Salem Limestone is to the sculptor's chisel.

Once the urban explorer learns to recognize the Salem, it's a constant companion through life, at least if one stays in America. In the heart of every city, in each of its suburbs, in one-stop-sign towns halfway to Santa Fe, there is always some of this utterly unassuming stone lurking. It's in the official face of the shoebox post office, on the Gothic ramparts of the local high school, in the lintels and window trim of the side-street Methodist church. And when you turn here to face south from the bridge down Michigan Avenue, it quickly becomes apparent how the Salem's gray-buff anonymity can also add an understated majesty to the lines of some of the finest twentieth-century skyscrapers.

5.2 London Guarantee & Accident Building

360 N. Michigan Avenue
Current name: LondonHouse Chicago Hotel
Completed in 1923
Architect: Alfred S. Alschuler
Foundation: Rock caissons
Geologic feature: Salem Limestone

This superbly sited building, with its elegantly curving elevation and monumental Corinthian columns, marks the spot where once Fort Dearborn stood. It harmonizes well with the much more modestly scaled Michigan Avenue Bridge Houses across the drive. For one thing, both sites reflect the Beaux Arts vision propounded by architects Daniel Burnham and Edward Bennett in their 1909 *Plan of Chicago*. For another, both feature an exterior of Salem Limestone, the widely used Mississippian-subperiod biocalcarenite that so splendidly shows off the stone carver's art. Here it's most readily apparent in the ornament of classical Greek meander frets that runs as a horizontal band atop the ground-floor windows.

While the Salem has a well-deserved reputation for good durability in most exterior settings, it, like every other building material one is likely to see in the city, is bound to eventually degrade in Chicago's lethal brew of precipitation, pollution, and temperature extremes. Here, after nine decades of exposure, it had suffered considerable crumbling and spalling—a fact that necessitated a major renovation conducted from 2014 to 2016. Some 5,000 square feet of new Salem were installed to replace the deteriorated cladding.

5.3   333 N. Michigan Avenue

Completed in 1928
Architectural firm: Holabird & Root
Foundation: Rock caissons
Geologic features: Salem Limestone, Morton Gneiss, Larissa Ophical-
   cite, Terra-Cotta, Cast Iron

Directly across Michigan Avenue from the London Guarantee Building stands this stylishly set-back skyscraper, which both compares and contrasts with the former. The comparison lies in its extensive use of Salem Limestone; the contrast, in its soaring, rectilinear Art Deco style. But that's not the only difference.

If you directly scrutinize the exterior of the lowest stories of 333—as I've always been compelled to do with my tour groups—you'll discover stone clad-ding at ground level that offers a story much different than the Salem's. This is our first manifestation of what I call the Grand Art Deco Formula: a huge expanse of the limestone above, literally a monolithic mass, footed by a crystalline rock type that is a darker, more detailed, and sometimes more subversive contrast to the beige uniformity above it. Here that crystalline rock is a riot of twisting and contorted forms in a jumble of swirling colors and shades. In one place on the building's western face there hovers within it a large angular mass of jet-colored amphibolite, sheathed in concentric rings of lighter-tinted minerals. Its trailing edge is pinched and pulled out into an attenuated tip, like a strand of sun-warmed saltwater taffy. Obviously, it's been subjected to immense forces and intense heat and pressure.

This splashy, wild-spirited rock, which can be found again and again on the Art Deco buildings of various Chicagoland neighborhoods and in Rockford and Milwaukee as well, was much beloved by designers of that era. Stone pro-ducers and marketers, with their unerring ear for the geologic misnomer, call it "Rainbow Granite" or "Oriental Granite," inadvertently suggesting it's purely and simply igneous. In fact, it's Morton Gneiss. The first part of its name honors the small southern Minnesota town from which it comes. The second, which my college Earth Science students relentlessly pronounced *guh-neese* until I told them it's actually *nice*, indicates that it's a foliated metamorphic rock. But spe-cialists in rock classification refer to the Morton stone even more precisely as a *migmatitic* gneiss—a mixed or hybrid metamorphic-igneous oddball that in this case is a crazy quilt of older gray gneiss derived from tonalite and granodiorite, black amphibolite that originally was komatiite or basalt, and another somewhat younger pink-and-white gneiss that started its career as pegmatite and quartz monzonite. As though that weren't enough to set the head spinning, it also con-tains aplite, a finer-grained granite. Realizing that such a complicated petrologic potpourri is liable to make the reader's eyes glaze over, I've developed my own less rigorous definition: *The Morton Gneiss is a rock with a very troubled history that in the course of eons has been messed up in amazingly beautiful ways.*

This flamboyant stone has been isotope-dated to the age of approximately 3.52 Ga by geochemists using the ultrasophisticated but risibly abbreviated SHRIMP technique. (It stands for sensitive high-resolution ion micro-probe.) This makes the Morton a representative of one of the oldest surviving rock units on Earth, and certainly the most ancient used in American architecture. And it also means that it's three-quarters the entire age of our planet, the sun, and our other solar system neighbors. When it formed, in far-distant Paleoarchean time, the most advanced types of life were rudimentary single-celled microbes; the Moon, less than half the distance from the Earth it now is, loomed large in the anoxic sky. I often tell my tour companions to run their fingers reverentially over the cladding here. Unless they chance upon a meteorite, this will be the oldest thing they'll touch in their entire fleeting lives. The sums of their existence, yours, and mine would not even be the slightest scratch on the Willis Tower geologic time line described in the first section of this book. But this stone's history would extend on that vertical scale from the sidewalk to the building's eighty-eighth floor.

FIGURE 5.2.   A fascinating jumble of serpentinite and green-haloed marble fragments, Greece's Larissa Ophicalcite was a favorite of Byzantine cathedral-builders a millennium and a half before it was used to adorn the Art Deco lobby of 333 North Michigan.

As if that weren't enough awe-inspiring geologic content for one building, 333's exterior also features handsome cast iron entrance and window ornament, and terra-cotta spandrels above the Morton cladding. Inside, the walls of the ground-floor lobby feature another exotic stone of tremendous visual impact and violent origin. This is the venerable Larissa Ophicalcite, a coarse-grained Hellenic sedimentary rock that had its origin in the Jurassic or Cretaceous. It's composed of hefty angular chunks of serpentinite and marble. These metamorphic fragments were broken apart by the frenetic activity of what is now Greece's fault-ridden crust, in the midst of the tectonic madhouse that the eastern Mediterranean was and still is.

On a recent visit here, I pointed out to the security guard on duty (who was warily watching me pay an abnormal amount of attention to the décor) the pale pieces of marble that float in the breccia's matrix like lonely marshmallows. These bright-white bits are mantled in thick jackets of the mint-green mineral serpentine. My new acquaintance found this particularly interesting, and before I knew it we were on a lobby-wide Easter-egg hunt for every example we could find. And so another inquisitive Chicagoan had joined the ranks of various guards, concierges, maintenance workers, bond traders, attorneys, tourists, and patrolling policemen I've met who were willing to join in some geologic discovery on their own turf.

In the stone trade of recent centuries, Larissa Ophicalcite was considered an exotic marble and was dubbed "Verde Antico." But it was first used in Imperial Roman and Byzantine times, when it was known simply as "Green Thessalian Stone." I first came upon this rock type not close to home here at 333 North Mich, but while standing in an advanced state of geologic astonishment amid a forest of massive columns made from it, in Istanbul's Hagia Sophia, the Church of Holy Wisdom. That incomparable place, with its interior domed heavenscape rivalled only by Rome's Pantheon, was completed in 537 CE in what was then Constantinople, at the express command of the emperor Justinian I.

5.4   Old Republic Building

307 N. Michigan Avenue
Former name: Bell Building
Completed in 1925
Architectural firm: K. M. Vitzhum & J. J. Burns
Foundation: Hardpan caissons
Geologic features: Deer Isle Granite, Northwestern Terra Cotta

This impressive building with its imposing three-story arched entrance is an essay in how to artfully and intelligently clothe a neoclassical high-rise. The bulk of the exterior is covered in white terra-cotta crafted by Chicago's own Northwestern Terra Cotta Company. It could pass at a distance for intricately carved marble, and is in fact a more enduring, less expensive, and more easily maintained alternative to it. Close inspection reveals how just above the base course the individual molded terra-cotta units have been assembled into running ornaments sporting botanical motifs.

The damp course, though, is real stone, and an excellent choice of one at that. It's polished Deer Isle Granite, quarried on tiny Crotch Island, just across the water from the town of Stonington, Maine, and down the coast a bit from Acadia National Park. This granite seems to have done a good job here shielding the base of the building from the onslaught of deicer salts in winter meltwater that can cause bad scaling and efflorescence, most notably in the Salem Limestone and even in other less porous stones, including some granites.

The geologist T. Nelson Dale, writing at about the same time this structure was designed, described the Deer Isle as "Lavender tinted medium gray. Coarse-grained, even-grained." To my eye, the gray has more of a beige tint, but then it must be noted that even stone taken from a single quarry can vary significantly in coloration. The type of surface finish it receives can also make a surprising difference. That said, here the stone provides a darkly handsome contrast to the light-toned terra-cotta above it.

The Deer Isle Granite is Devonian in age and formed as part of a swarm of giant blobs of magma that rose from the Earth's interior when the microcontinent that geologists call Avalonia merged with part of the ancestral edge of North America. That collision, and others caused the arrival of other crustal fragments, created the once lofty but now long-vanished range of the Acadian Mountains. While the magma body containing the Deer Isle Granite did not reach the surface before it solidified, it was eventually exposed there by erosion. And then, in the long course of time, it was attacked by yet another formidable force, human extraction technology powered by the profit motive. The ultimate result of that is the adornment of a beautiful building with this handsome stone.

5.5 Carbide & Carbon Building

230 N. Michigan Avenue
Current name: St. Jane Hotel
Completed in 1929
Architectural firm: Holabird & Root
Foundation: Rock caissons
Geologic features: Mellen Gabbro, Holston Limestone, Belgian Black
  Limestone, Northwestern Terra Cotta

While the nearby 333 North Michigan has a secure place in the Windy City's Art Deco roll of honor, the Carbide & Carbon is probably the one most praised example of this style in this section of the Loop. And justly so. It's a glorious essay in one-and-onliness that abrogates the standard skyscraper design formula of its era: there's none of the usual bland Salem Limestone to be seen anywhere on it. Completed in the last, economically disastrous year of the Roaring Twenties, this slender beauty is decked out from the fourth story up in handsome gold and dark-green cladding crafted by the Northwestern Terra Cotta Company. Below that, a glistening stone builders call "black granite." These two classy materials complement each other perfectly. But here the geologic explorer is confronted with a "granite" that really isn't. While this rock type does have an igneous origin, it occupies its own position quite apart from granite on the geology textbook's identification chart. One can choose not to care much about this lack of terminological accuracy. And by the same token, I suppose, one can call a native of China a Swede—the species identification is correct, after all—but in doing so one rather badly misrepresents the person's origin and basic identity.

As it is, the architect's "black granite" tag covers quite an assortment of darkly appealing rock types—coarse-grained gabbros and anorthosites, medium-grained diabases, and only rarely a stone that is indeed a true granite, or at least a granitoid. In the case of the Carbide & Carbon, the cladding is the Mellen Gabbro, once quarried in the North Woods Wisconsin town of that name, some thirty miles south of the Apostle Islands. In the early twentieth century this stone went by various trade names, including the tautological "Veined Ebony Black" (the variety used here), "Primax Granite," and "Rosetta Black." And though it may never have been as famous as the real Rosetta stone, it was at one time a popular choice of Chicago architects.

While it can't compete with the extremely ancient age or undulating patterns of 333 North Michigan's Morton Gneiss, the finely polished Mellen here on display is both aesthetically appealing and scientifically intriguing. Composed of large crystals of plagioclase feldspar and other dark silicate minerals, including some that have the glittering reflective property called labradorescence, it dates to the Mesoproterozoic era, a time when northern Wisconsin experienced a great deal of igneous activity and upwelling of magma associated with the great Mid-continent Rift event. From about 1.1 to 1.0 Ga this massive tectonic upheaval threatened to tear the ancestral form of North America apart.

The Carbide & Carbon's geologic bounty is not confined to its exterior, as a quick dash in through the Michigan Avenue entrance reveals. Here the rockhound's gaze is immediately drawn downward. No expense was spared with the flooring: the light brownish-gray surface there is the Ordovician-period Holston Limestone, marketed to this day as "Tennessee Marble." Quarried in a long and narrow band from the northeast to the southwest of Knoxville, it graciously accepts the high-gloss polish that a true metamorphosed marble characteristically does. One of the Holston's primary identification traits, which can be spotted in many of the region's building interiors, is its tendency to sport long, thin, tightly crinkled lines called stylolites. They're definitely present here.

And bordering the Holston on the sides of the floor is trim of white-veined Belgian Black Limestone, quarried from the Paleozoic strata of Namur and other localities in that Low Country. Not surprisingly, it's often called a marble, too, though like the Holston it was never subjected to the intense crustal pressures or heightened temperatures that recrystallize limestone into real, sugary-textured marble. Be that as it is, the Belgian Black and the Holston, so geologically similar but so different in aspect, give some indication of the great range of colors to be found in commercially quarried limestones. In fact, the Holston is polychromatic all by itself, with other forms that are pale to rose pink or even a ruddy coral red with gray flecking and white veins.

5.6   Seventeenth Church of Christ, Scientist

55 E. Wacker Drive
Completed in 1968
Architectural firm: Harry Weese & Associates
Geologic feature: Tivoli Travertine

This squat and stolid structure might look hopelessly lost among its lofty neighbors were it not for its distinctive semicircular bulk, which resembles the prow of a broad-beamed supertanker nosing its way into the intersection. The church's exterior cladding is its geologic highlight. What might strike passersby as concrete—not a bad guess, given its popularity with modernist designers—is actually one of the most famous building stones in the history of architecture. At a distance it seems to be a smooth, off-white surface, but it's actually pitted with small cavities. This is Tivoli Travertine, a favorite of ancient Roman architects, who used it in a variety of temples, the Forum of Augustus, and that most famous Eternal City landmark, the Colosseum. Also known as "Roman Travertine" and "Italian Travertine," this rock type is still extensively quarried east of Rome, mostly in the Acque Albule basin near the historic town of Tivoli.

Travertine is an unusual form of limestone that owes its origin to the existence of groundwater springs, sometimes hot and sometimes cool, that produce water rich in calcium compounds. Often the presence of these springs is linked to local volcanic activity, and indeed the Tivoli region has been the scene of a great deal of volcanism in the recent geologic past. Over time calcium carbonate (otherwise known as calcite, limestone's primary mineral), precipitates out of the water issuing from a spring to form a deposit of layered stone that is pitted or even spongy in texture, if the spring's flow rate is especially brisk. The ornamental appeal of the Tivoli Travertine lies in its irregularly striped or banded patterns, and in the ease of its cutting and carving. While the banding was originally oriented more or less horizontally, sections of the stone are sometimes mounted so the pattern is vertical, as appears to be the case for at least some of the cladding here.

This building stone also makes a dramatic temporal as well as visual contrast to the Morton Gneiss on display just a block away on the lower façade of 333 North Michigan. At the hoary age of 3.52 Ga, the Morton is the most ancient rock type on display in Chicago's architecture; the Tivoli, most likely the youngest. The Acque Albule strata from which it comes are Pleistocene to Holocene in age, a mere 165 ka or younger. To the geologist that's just the merest flick of an eyelash in time.

5.7 1 Prudential Plaza

130 E. Randolph Street
Former name: Prudential Building
Completed in 1954
Architectural firm: Naess & Murphy
Foundation: Rock caissons
Geologic features: Salem Limestone, Støren Trondhjemite, Mondariz
    Granite, Aluminum

I remember the day when, as a young child on a visit downtown, my father pointed out to me the newly erected Prudential Building, as it was then called. He proudly noted that it was now Chicago's tallest, and the latest thing in modern architecture. The first skyscraper completed in the city since the Great Depression, it seemed to loom over everything as the harbinger of a high-tech future. That said, it was also one of the last Chicago high-rises anchored with rock caissons laboriously dug by hand. Because it was built over the Illinois Central Railroad tracks, its stable foundation was connected to the visible part of the edifice with a subterranean maze of 500 stilts.

One has only to look around the site to see what has happened since the middle of the twentieth century. A clump of taller towers has sprung up all around it, like a patch of weed trees high on excess soil nitrogen. Still, in a geologic sense, 1 Prudential Plaza—now nicknamed One Pru as a complement to the more euphonious Two Pru of its younger neighbor next door—preserves its special significance. It also serves as an excellent case study in how the geologic sleuth must often do a good deal of patient digging, sifting, and checking on-site to discover the real identity of the stones that bedeck Windy City structures.

The most extensively used rock type here at One Pru is the one most extensively used in Chicago generally. On the building's principal, south-facing exposure, it can be seen at height as narrow piers running between the windows with their spandrels of corrugated aluminum. On the west-facing surfaces it forms broader sections on the upper portions. But on the blanker northern wall, facing Pru Two's courtyard, it extends all the way to ground level. There you can easily study its beige, dressed-face surface. Its porous and finely granular texture, its telltale fossil-shell fragments, and its instant reactivity to a drop of dilute hydrochloric acid identify it at once as the famous Salem Limestone, a biocalcarenite of Mississippian age more familiar to most as "Indiana Limestone" or "Bedford Stone."

In the late 1990s a rehabilitation project following the construction of Two Pru resulted in the lower piers and some of the ground-floor cladding of One Pru being redone in pinkish-gray Mondariz Granite and other stone types as well. Interestingly, a December 1996 article in the *Chicago Tribune* noted that the building's managers had become justifiably concerned by the rusting of some of the Salem's steel fasteners. Fearing that falling stone slabs might leave too deep an impression on passing cars and pedestrians, they'd decided to replace all the Salem with the similarly colored Mondariz. On top of that, a brown granite from Canada would take the place of the aluminum panels. However, a quick stroll around One Pru demonstrates that this total makeover didn't actually happen. But a partial one did.

Nowadays on One Pru's south-oriented façade you'll see the following sequence from bottom to top: a thin basal trim of an unrecorded black igneous rock, then what appears to be a darker-finished version of the Mondariz Granite, then narrow trim of an unidentified red granite, and then Mondariz of a considerably lighter finish extending to the top of the grand entranceway windows. All these stone varieties, known and unknown, apparently date to the late 1990s renovation. But above them, the materials originally chosen, the Salem Limestone and the aluminum spandrels, remain.

While the origin and composition of the Mondariz Granite is covered in the Two Pru description, the aluminum is a unique design element of One Pru and therefore deserves to be recognized here. After all, it's as geologically derived as any building stone is. Indeed, it is our planet's most abundant metal, and third most common element, though it's thought to be almost completely concentrated in the uppermost 10 miles of the Earth's crust. Now found in a multitude of human applications, it is chiefly extracted from bauxite, an unusual ore rock that forms in wet, tropical climates. Chicago architects started using aluminum for exterior ornament in the twentieth century.

The foregoing inventory of stone and metal demonstrates that One Pru is a building of considerable geologic diversity. Yet it offers one additional treasure, and to any geologist it would have to be the greatest of the lot. I'd seen a reference to a "Norway White" granite used as cladding here, but I couldn't pin down its identity till I stumbled across another and much older *Tribune* article, from May 1954. It reported that "twenty-three men and three women from a Norwegian ship" were offloading a special cargo on the south bank of the Chicago River, just a little lakeward of the Michigan Avenue Bridge. It was "525 tons of Stern silver-white granite" that would be cut into "3.5-inch slabs" and then installed on the newly minted Prudential Building's ground-floor exterior.

While I could find no record of a Stern granite quarried then or now in the land of the fjords, I discovered there is one that hails from the town of Støren—which, when spoken by a native Norwegian, sounds very much like what the anglophone American ear would hear as "Stern." I can easily picture a moderately bored city-desk *Trib* reporter out on this routine assignment, asking a crewman from the ship for the name of the cargo, and then distractedly jotting down the phonetic equivalent of what was actually spoken. And it turns out that among the Støren stones various trade names one can find both "Støren Silver White" and "Norway White."

When I contacted the Norwegian Geological Survey for more information, I learned the rock was actually a rare granitoid type known as trondhjemite (TRON-yem-ite). What distinguishes it from normal granite is that, instead of having fairly equal amounts of potassium-feldspar and plagioclase-feldspar minerals, it contains almost nothing but the latter, specifically in the form of oligoclase. The Støren Trondhjemite has been dated to approximately 432 Ma, which means it's Silurian. It formed as a rising body of magma during the Caledonian Orogeny, a mountain-building event caused by the collision of Baltica, the ancestral version of northern Europe, and the precursor of North America, Laurentia.

Fortunately, this rarest component of the original Prudential Building design is still present in places. Look for it at the base of the west-facing wall that bears the large Rock of Gibraltar logo. Farther up, it's Salem, but the first 15 feet or so, on either side of the restaurant entrance, are clad in the Støren. It's also still extant in the planters that line the Randolph Street sidewalk. You'll be able to identify it as the medium-grained salt-and-pepper stone that lacks the Mondariz Granite's overall beige tint and large pink feldspar crystals. Some of the stone planter sections show a distinctly striped pattern known as flow banding, which is often a diagnostic feature of this rock.

FIGURE 5.3. The Støren Trondhjemite that survives at 1 Prudential Plaza serves as both exterior wall cladding and in planter rims, as here, where it's garnished with a helping of ornamental cabbage and mini-mums. In this spot especially its tendency to exhibit flow banding is easily discernible.

5.8   2 Prudential Plaza

180 N. Stetson Avenue
Completed in 1990
Architectural firm: Loebl Schlossman & Hackl
Geologic feature: Mondariz Granite

Various comparisons to a pulp-sci-fi rocket ship notwithstanding, this building reminds me most of a tall-bladed and complexly faceted hornblende crystal, especially when I lean back and look up the full height of its western exposure, darkened by the shade of its hemmed-in site. Of the assortment of igneous intrusive rock types that have been used for the exterior's trim and cladding, apparently only one has been documented by specific name.

Happily, the rock type that can be identified is well worth a visit. Both in the plaza and on Two Pru's façade piers, the handsome stone selected is the coarse-grained, pinkish-gray Mondariz Granite. It comes from Ponteareas in northwestern Spain, just a few miles north of the Portuguese border. Like the other granite selections produced in this major quarrying region, the Mondariz owes its existence to the Variscan Orogeny, the great European mountain-building event that occurred during the assembly of the supercontinent Pangaea. In the Carboniferous period the process of plate subduction generated numerous masses of magma, derived from sinking ocean crust and sediments that rose and eventually solidified while still underground. At Two Pru the Mondariz is present in different finishes that impart lighter and darker shades to it, but never quite as dark as the unidentified brown granite, apparently from Canada, that makes up the cladding between the piers. Note how coarse the texture of the Mondariz Granite is—this is an indication that its magma took a very long time to cool underground. Also take a close look at its big pink feldspar crystals, which give the rock its overall tint.

5.9   Aon Center

200 E. Randolph Street
Former names: Standard Oil Building, Amoco Building
Completed in 1973
Architect: Edward Durell Stone
Geologic features: Weathering of the original Carrara Marble, Mount
    Airy Granodiorite, Beryllium-Copper Alloy, Naval Brass

No other building in America has had more ink spilt about its cladding. And for good reason; accounts of the dismal and expensive failure of this imposing tower's original exterior stone often read like morality plays decrying corporate hubris and the folly of aesthetic whimsy overriding sober engineering savvy. The fact remains, though, that this building's initial appearance when it was clothed in the whiteness of premium-grade Carrara Marble was an opalescent and awe-inspiring sight, especially when beheld near sunrise or sunset. But that beauty ultimately came at much too great a cost. When in 1992 a handsome but admittedly duller American granitoid had completely replaced the gleaming Italian marble, estimates of the makeover bill ranged from 60 to 80 million dollars. This has been described rather melodramatically as being half or even more than the price of constructing the whole building in the first place. But when one makes the effort to factor in inflation and the downward slide of the dollar's buying power from 1973 to 1992, one finds the figure is at most a little less than one-quarter the original cost. Still, that's a whopper of a bill.

Problems with the Carrara began not long after this dramatically sited "super-tall" (sometimes still called by its early Standard-Oil-phase nickname of Big Stan) was formally opened for business. Some of the marble panels were flung off the building by wind shear; and by the end of the 1970s inspections of the exterior had revealed both cracking and warping of the stone. By 1988, almost a third of the cladding slabs were bowed outward at least half an inch, and sometimes three times as much as that. Stopgap securing straps were added, but it soon became clear the only answer would be to remove the Carrara completely and replace it with a more enduring rock type.

This colossal replacement job was the sad result of initially using a porous ornamental stone much better suited for bank lobbies and art galleries than for outer surfaces exposed to the harsh realities of Chicago's unforgiving continental climate. There was the onslaught of the rain, naturally acidic and potentially even more so due to pollutants, that generated microscopic cracks in the marble. Then there was the relentless diurnal cycle of thermal expansion and contraction, and the additional invasion of atmospheric moisture into the marble's pore spaces. As though nature's coordinated, multipronged attack wasn't enough, there was also the unlucky fact that the Carrara panels were probably just too thin in any case. Recently invented stone-processing technology had made it possible to slice the Carrara panels to just 1.25 to 1.5 inches thick. This was a cost-cutting measure that the Standard Oil of Indiana executives had surely appreciated, but it proved to be an unwitting exercise in penny-wisdom and pound-foolishness. In the end, according to civil engineers Matthys Levy and Mario Salvadori, writing in their morbidly fascinating study *Why Buildings Fall Down*, the slimness of the cladding was indeed "a major contributory factor" to its deterioration.

The Aon Center's troubles are by no means the only example of marble's unsuitability on high-rise exteriors in humid, extreme-temperature climates. In the case of the Carrara Marble the first highly publicized American fiasco was not Big Stan but the First Lincoln Tower in Rochester, New York. Since then both Italian marbles and those produced elsewhere have suffered warping and other forms of damage in various locations across the world.

Fortunately for the urban geologist, the Aon Center is still worth a lingering visit, and it's one I always include in my downtown tours. The first thing my fellow travelers notice on arrival here is what remains of Harry Bertoia's visually intriguing and sonically enthralling, if untitled, "sounding sculpture." It's a modernist's Aeolian harp, and it originally featured eleven clusters of beryllium-copper alloy rods of different heights welded at their base to naval brass plates. Now only six remain. This work of kinetic art is a geologic point of interest all in itself because of its metal content. Beryllium-copper, dubbed BeCu in honor of its elements' chemical symbols, is an alloy widely used for electronic connectors and computer components. While copper, nowadays derived from such sulfide minerals as chalcocite and chalcopyrite, has a record of human use ranging back to at least 8,700 BCE, the rarer constituent, beryllium, is a much more recent discovery. No doubt that's partly due to its being merely the forty-fourth most common element in the Earth's crust. It's extracted from the mineral bertrandite and from beryl ore, substances currently mined only in the US, China, and Kazakhstan. And the naval brass of the bases is in turn an alloy of copper, tin, and zinc renowned for its strength and resistance to corrosion.

Back in the realm of stone, the Aon Center's replacement cladding may lack the instant touristic appeal of the sounding sculpture, but on closer inspection you'll discover it's an excellent place to scrutinize a fascinating American granitoid rock type. The best place to see it is along the walkway of the shaded plaza by the building's western face. This is the Mount Airy Granodiorite, quarried in the town of the same name in North Carolina's Piedmont. Better known to architects as "White Mount Airy Granite," geologists classify it as a granodiorite instead because a greater percentage of its total feldspar content belongs to the plagioclase group.

From some distance the Mount Airy gives the hulking skyscraper a flat-white appearance; up close it reveals itself to be a medium-grained, salt-and-pepper menagerie of minerals: white to beige feldspars, gray quartz, and the black mica known as biotite. These crystals solidified in a pluton, a large mass of magma that did not reach the surface before it cooled some 335 million years ago. So this stone dates from the Mississippian subperiod, a much earlier origin than that of the Carrara, which began as Jurassic-period limestone that wasn't metamorphosed into marble until the Oligocene epoch.

The panels here were wisely cut two inches deep. While the Notre Dame University geologist and building-stone expert Erhard Winkler expressed concern that even they too could conceivably buckle eventually—even though granite lacks marble's calcium-carbonate content and subsequent vulnerability to acidic rain—Levy and Salvadori suggest that the stouter thickness should prevent that from ever occurring. It has now been almost three decades since the new cladding was installed, and as far as I've been able to discern, it's a case of So Far, So Good.

5.10 Chicago Cultural Center

78 E. Washington Street
Former name: Chicago Public Library (main branch)
Completed in 1897
Architectural firm: Shepley, Rutan & Coolidge
Foundation: Wooden piles
Geologic features: Salem Limestone, Stone Mountain Granite, Carrara Marble, Connemara Marble, Shelburne Marble, Vermont Serpentinite, Favrile Glass

In *The Brothers Karamazov* Fyodor Dostoyevsky wrote that "some good, sacred memory, preserved from childhood, is perhaps the best education." Surely I'm not the only person left who has a sacred memory of this noble Italian Renaissance Revival edifice from its earlier existence as Chicago's flagship library. To this day I recall one golden afternoon when a much younger version of myself sat in its decorously appointed Reading Room, ostensibly researching a high school English paper but mostly gazing out at the sunlit hustle and bustle of the street below.

Now that the Harold Washington Library Center on South State Street has taken over this structure's original role, the Cultural Center functions as a facility for public lectures and meetings, casual lunchtime get-togethers, seniors' Scrabble clubs, art installations, museum exhibits, private events, and more. It's also the best one-stop geology lesson to be had in the Loop. I always start my downtown walking tours here. But the building is so rich in talking points that it's difficult to get back out the door to see the rest of the sights.

So perhaps the best place to start this geology (and here also botany) lesson is on the outside. Below this great anthropogenic rock outcrop of a building sits a pile foundation that is, in effect, a subterranean plantation of *Pinus resinosa*, the admirably straight-trunked red pine. A species long used for piling and utility poles, it's a certified North American native that is often known instead by the misleading nickname of Norway pine. This underground stabilization system was designed by the renowned civil engineer William Sooy Smith. As the *American Architect* reported in 1893, the job required a 4-ton steam-engine rig with a 4,500-pound hammer that delivered 54 bone-rattling blows per minute till each 50-foot pine trunk reached the hardpan. Once all the piles were in place Smith decided to cut their tops off 14 feet below the "city datum"—a benchmark set approximately at the level of Lake Michigan—and use stone-rubble caps for the remainder up to grade. He feared that if left in place, the uppermost sections of the timber would sit above the water table—the upper boundary of the groundwater—and thus be more liable to rot. This appears to have been an unfounded concern, but his meticulousness was typical of how this building was constructed.

The exterior is primarily clad in Salem Limestone, the architect's "Bedford Stone" or "Indiana Limestone," and so sustains a dominant leitmotif along Michigan Avenue and practically everywhere else in town, too. And, it must be said, it's the perfect choice here. Its unassuming light-gray sobriety is a fine match for the rows of arched windows and Ionic columns. The ashlar blocks are dressed-face, which is to say their outer surfaces are smooth and planar, like oversized concrete blocks. Look at this Mississippian-subperiod sedimentary rock carefully, and preferably with a hand lens, and you'll quickly spot a multitude of tiny fossil fragments.

Beneath the Salem lies a damp course of another gray rock type of slightly darker tone. This is the Stone Mountain Granite, quite a rarity in Chicago nowadays but a wise choice for this part of the exterior that is in direct contact with the sidewalk and its nasty goulash of dirty slush, saline meltwater, discarded chewing gum, soot, spilled paint, and a depressingly large inventory of urban unmentionables. Had the Cultural Center's designers done what some other less experienced architects have, and continued the Salem all the way down to grade, there probably would have been geochemical hell to pay: pronounced pitting, exfoliation, salt-crystal efflorescence, and excessive case hardening of that more porous and vulnerable rock. Generally granite holds up considerably better, though one can see here that even the Stone Mountain has taken its share of affronts over the past twelve decades. Made of small interlocking mineral crystals, some of which are more susceptible to weathering than others, granite can crack, spall, warp, and become infiltrated by that wiliest of compounds, water. Here there's also evidence of invasion by dissolved calcite leached out of the Salem above. As with any substance or entity exposed at the Earth's surface, it's only a matter of time before everything is reduced to dust or rubble. But more often than not granite damp courses fight the best delaying action possible.

The Stone Mountain Granite is named for the famous mass of igneous rock, now a park, fifteen miles east of Atlanta, Georgia. In its native habitat—for example, at the summit of Stone Mountain itself—the rock often contains "cat's paws," groups of large dark tourmaline phenocrysts set within pale haloes, but the stone I've seen here seems to have been selected by its quarriers, the Venable Brothers Company, for its gray uniformity. Isotopically dated to 291 Ma, in the lower Permian period, it first formed as a pluton deep underground and was ultimately exposed by the implacable forces of erosion.

To see the splendors of the Cultural Center's interior, the best place to start is at the building's southern, Washington Street end. Once through the doors the visitor is almost overwhelmed with the lobby's opulence. This can elicit a reaction I've heard more than once: *They don't make public libraries like this anymore.* The grand staircase that rises in the center gleams in white Carrara Marble, apparently of the tiptop Statuario grade, as an 1897 newspaper account of the newly built library strongly suggests. This was, as every chronicler of the Carrara invariably mentions, Michelangelo's preferred carving stone, and the great sculptor's travails in getting an ample supply of it for his own projects is the stuff of legend. A letter he wrote to his brother in 1518 brims over with bile at the logistical problems he encountered:

> I shall mount horse at once and go to find Cardinal Medici and the Pope and tell them my situation, and leave the project and go back to Carrara,

which they pray me to do as one prays to Christ. These stonecutters I brought down . . . don't understand the first thing about quarries or marbles either. They have already cost me more than a hundred and thirty ducats and they have not yet quarried me a slab of marble that is any good, and they go about faking that they have found great things and try to work for the Board of Works and others with the money they've received from me. (Creighton Gilbert translation, by permission)

Nor was Michelangelo the only one to ever kvetch about what a Carrara project can cost in both monetary and emotional capital. After all, this is the stone that the Aon Center's exterior was originally and disastrously clad in. But an interior setting such as the Cultural Center's is exactly where it should be used. Here one can fully comprehend the marble's sparkling worldwide reputation. It has transformed this structure into a temple of learning.

The marble's attractiveness has been further enhanced with insets of varicolored Favrile Glass—in effect, elaborate geometrical patterns of silica set in a sea of calcite. Favrile Glass was named, developed, and patented by Louis Comfort Tiffany (1848–1933), and most sources just refer to this opalescent ornamental material as Tiffany Glass. Nestled within it and the marble on the panels of the staircase handrails one can also find disks of sea-green Connemara Marble. Its color is wholly appropriate, given that it was quarried in County Galway, Ireland. This fetching stone is an unusual form of marble that is rich in the green mineral serpentine. The Connemara originated as limestone in the upper Neoproterozoic era and was metamorphosed into its present form in the Ordovician period. This makes it considerably older than the Carrara, which began as limestone in the lower Jurassic period and did not get heated and compressed into marble until the Oligocene epoch.

A long passageway, itself an exhibit gallery, stands to the left as you face the staircase. It connects the Washington Street lobby with the building's northern end, where you'll emerge first into the public seating area now called Randolph Square. It too features Carrara Marble as wall cladding, with pilasters and trim of nicely contrasting beige Holston Limestone, otherwise known as "Tennessee Marble." The Holston is close to being the interior-space analog of the Salem Limestone, in that it's found in a multitude of architectural settings across America. But also like the Salem, it should never be considered a petrological cliché because of its popularity, which is resoundingly justified.

The lobby just beyond Randolph Square features an information desk and more marble wall cladding. Here, however, it's not Carrara but the equally splendid Shelburne, a lower-Ordovician true American marble quarried along Vermont's western flank. While available records do not indicate where exactly the stone here was quarried, it's most likely from the West Rutland or Dorset areas. In any case, it's a white and green-veined variety that resembles the type known in the trade as "Cipollino"—the Italian word, believe it or not, for "baby onion." This, a time-honored term of the scalpellini (Italian stonecutters), refers to the fact that the marble's ply-upon-ply patterning resembles the exposed layers of a cut onion. Or at least it does if one is sufficiently imaginative. Here the Shelburne Marble has been mounted with its striping oriented more or less horizontally.

The hall at the southern end of Randolph Square and the staircase within it both feature more Holston Limestone. Once up the stairs and on the second floor, you'll enter yet another Cultural Center showplace, the Grand Army of the Republic (G.A.R.) Hall and its adjoining spaces. The antechamber has some of the best Holston panels of all. Note the bold series of stylolites, the jaggedly crinkled veins that march across them like a procession of lightning bolts. This is a classic distinguishing feature of this stone.

The G.A.R. Hall itself is a geologic tour de force. The wall cladding here is nothing short of breathtaking: panels of darkest green merging into Stygian blackness in the spots not hit by the windows' shafts of light. It would all be overwhelmingly somber were it not for the stone's highly reflective polish and the webbing of white and light-green veins crisscrossing the panels. The rock's primary mineral constituent, serpentine, accounts for the deep-green base color, as it does for the Connemara Marble described above. The lighter veins are magnesite, a relatively uncommon, magnesium-containing carbonate.

In the building and stone trades this striking stone goes by the sesquipedalian name of Vermont Verde Antique Marble (or Serpentine), a farrago of English, Italian, and French no doubt concocted to dazzle those linguistically challenged enough to assume it actually makes sense to aesthetes. I've pared down that pastiche of transcultural windbaggery a bit, into the simpler and more geologically accurate Vermont Serpentinite. Like the Carrara and Shelburne Marbles, this rock type is metamorphic, but its origin is considerably more exotic. Long before the Atlantic Ocean existed, what we now call America's Northeast faced an earlier ocean, the Iapetus. In the late Neoproterozoic and early Cambrian, a type of ultramafic rock called dunite formed in the uppermost mantle beneath the ocean's crust. Later, in the Ordovician, an arc of volcanic islands, moving slowly but relentlessly toward ancestral New England, scraped up the dunite and other rock types on the seafloor and below it, and ultimately plastered them onto the margin of North America. The heat and pressure generated in the collision resulted in the dunite's first pulse of metamorphism. Then in the Devonian period the second and final transformation into serpentinite occurred when the minicontinent of Avalonia, following the fate of the island arc, plowed in behind it to produce the plate-tectonics equivalent of a three-car pileup. So this Vermont Serpentinite was brought to us for our viewing pleasure from a most inaccessible place, deep in our planet's interior. Along its journey of thousands of miles and hundreds of millions of years it was forged into its current form by two great events in the ever-evolving architectural renovation project that is our planet.

5.11   Chicago Athletic Association Annex

71 E. Madison Street
Completed in 1907
Architects: Richard E. Schmidt, Garden & Martin
Foundation: Rock caissons
Geologic features: Conway Granite, Salem Limestone

This slender structure tucked between stouter neighbors is in fact a side-street addition to the much grander Chicago Athletic Association Building of Venetian Gothic design fronting Michigan Avenue nearby. While the Annex's lower two stories are clad in Salem Limestone, complete with intricate carvings perched over the entrance, the real point of interest, definitely worthy of a quick stop, is the protective damp course below it. The builders wisely chose an enduring rock type to protect the Salem from deicer salts and water infiltration. This is the Conway Granite, quarried in Conway and Madison, New Hampshire. It is a coarse-grained igneous rock that owes its overall light-pink cast to orthoclase, its main alkali-feldspar mineral. Closer inspection also reveals a white plagioclase feldspar, black biotite mica, and a smoky quartz of grayish-brown or even purplish color.

The Conway is Jurassic in age, and formed at a time when northern New England was experiencing intense magmatic activity, including massive volcanic eruptions, during the breakup of the supercontinent of Pangaea. This particular granite formed from one of a swarm of felsic-magma masses that rose upward from the Earth's interior but cooled and solidified before reaching the surface. Today the landscapes of New Hampshire and neighboring Atlantic Seaboard states are studded with large areas of outcropping granites that are the exposed remnants of these plutons, as well as those of earlier geologic periods and other tectonic upheavals. That accounts for a great deal of ruggedly beautiful scenery, but it can also be a matter of concern for homeowners in northern New England because granites naturally contain isotopes that can produce carcinogenic radon gas, which has a tendency to accumulate especially in basements. In particular, the Conway Granite group, including both the "Conway Pink" and "Redstone Green" varieties, has been cited as possessing a relatively high level of radioactive thorium. But, to quote from a current US Environmental Protection Agency fact sheet, "Radon originating in the soil beneath homes is a more common problem and a far larger public health risk than radon from granite building materials." So the stone here is certainly safe to scrutinize, even if you choose to do so for more than a moment or two.

5.12 Willoughby Tower

8 S. Michigan Avenue
Completed in 1929
Architectural firm: Samuel N. Crowen & Associates
Foundation: Caissons (type not specified)
Geologic features: St. Cloud Area Granite, Salem Limestone

Situated on the southwestern corner of Michigan and Madison, this building features a double-doored main entrance framed in carved Salem Limestone. But the real geologic standout of its exterior is the cladding of the lowest two floors. It makes Willoughby Tower into a showplace for the "Rockville" variety of St. Cloud Area Granite quarried in central Minnesota. Isotopically dated to approximately 1.78 Ga and hence to the Paleoproterozoic, this stone is one of the most striking of all the true granites seen in the city, and it's been used widely here. It is very coarse-grained, with sharp-edged crystals up to three-quarters of an inch long. The primary mineral constituents are pink and white feldspars (mostly orthoclase, with some microcline and a little plagioclase), glassy gray quartz, and black biotite mica. The hefty size of these crystals indicates that the magma from which they formed cooled very slowly within the Earth's crust.

5.13 Gage Building

18 S. Michigan Avenue
Completed in 1899
Architectural firm: Holabird & Roche
Foundation: Wooden piles
Geologic features: Northwestern Terra Cotta, Boston Valley Terra Cotta

Nowadays the Gage Building is not so much a site for up-close inspection as it is for a summary view. It's best appreciated from the opposite side of Michigan Avenue, where its tall and slender exterior of white terra-cotta can be appreciated in a single lingering glance. Here the botanically inspired genius of Louis Sullivan, the façade's designer, is in full leaf and fruit, with the ornament culminating at the top in a pair of spectacular cartouches. They, the cladding units, and the other decorative details were originally fabricated by Chicago's Northwestern Terra Cotta Company, but in recent years many of the sections have been replaced by faithful replicas, including those skillfully duplicated, using clay mined in the Ohio River Valley, by the Boston Valley Terra Cotta works in western New York. Whatever the source of the clay for the terra-cotta now extant on the building—whether it was dug from a bed of Pennsylvanian shale in Illinois coal country or mined in a river floodplain hundreds of miles away—the fact remains that when molded, glazed, and fired it has transcended its humble origin and become a beautiful stonelike material marvelously adaptable to both the builder's practical needs and

FIGURE 5.4.   One of Louis Sullivan's original cartouches, painstakingly removed from the Gage Building's façade, reassembled on the factory floor of western New York's Boston Valley Terra Cotta works for comparison with its exact-duplicate replacement, still being assembled (top). Note how this intricate design, which from a distance appears to be one solid piece of sculpture, is actually composed of many individual cladding units that must be precisely fitted together. (Reproduced by permission of Boston Valley Terra Cotta)

the architect's artistic imagination. (Incidentally, at time of writing the left-hand or southern cartouche is the original Northwestern casting; the right-hand or northern is Boston Valley's.)

The Gage Building also once featured a marvelous ground-story front featuring more intricate Sullivan designed forms brilliantly realized in cast iron. Sadly, it was removed in the 1950s. The record of terra-cotta replacement here indicates that even an enduring material is ultimately subject to the onslaught of the elements. But the loss of the priceless cast iron ornament reveals something even more sobering: in Chicago the most brutal erosional force is humanity itself.

5.14   University Club

76 E. Monroe Street
Completed in 1909
Architectural firm: Holabird & Roche
Foundation: Wooden piles
Geologic features: Salem Limestone

The design ethic here, often described as Tudor Gothic or neo-Gothic, seems to suggest a reference to the once-prevailing look of America's most prestigious academic institutions. The exterior presents the urban geologist wandering up and down Michigan Avenue with an acute case of déjà vu: it's yet another massive outcropping, so to speak, of Salem Limestone, otherwise known as "Indiana Limestone" or "Bedford Stone." Since the Salem is directly accessible at ground-floor level, this is a good place to pull out one's hand lens (that indispensable companion of every roving rock hunter) and search for small fragments of Mississippian invertebrate fossils embedded in the ashlar's calcite matrix. These remnants of a long-vanished tropical sea are the helpful telltales that confirm this is indeed real limestone and not its anthropogenic look-alike, fine concrete.

By the time the University Club was erected, the Salem had already become a Windy City commonplace. It appears it was first introduced here not long after the Great Fire of 1871 by Chicagoan John Rawle, who'd visited southern Indiana and been so impressed by the Salem's superior qualities that he opened a stone yard on his return. And the city, in the midst of its frantic post-fire building boom, was an eager consumer of the stone even before it became all the rage across the rest of the country. Nor did contemporary geologists fail to sing its praises. Building-stone expert George Perkins Merrill, writing in 1891, noted that "the stone is soft, but tenacious (specimens having borne a pressure of 12,000 pounds per square inch), and works readily in every direction."

5.15 Crown Fountain

Millennium Park, just east of Michigan Avenue between Madison and
  Monroe Streets
Completed in 2004
Architectural firm: Kruek & Sexton
Artist: Jaume Plensa
Geologic features: Mashonaland Diabase

This popular park site brings welcome relief from the hubbub of Michigan Avenue. Two high towers, unnervingly reminiscent of the inscrutable alien monoliths of *2001: A Space Odyssey*, loom over a flat expanse of stone, gray when dry and inky black when wet. The towers may not teach park visitors how to take the next step in human evolution, but when weather conditions permit they do shoot out tumbling cascades and even long, arcing jets of water from the pursed lips of giant Chicagoan faces projected on their LED screens.

The fountain's dark flooring is composed of what architects and suppliers feel compelled to call "black granite." Whenever this term is bandied about, the geological explorer should be on high Misnomer Alert: "black granite" often isn't a granite. Usually, it's one of a number of other igneous rocks of deeper shade and a different chemistry. In the case of the Crown Fountain pavers, it's what American geologists call diabase—a mafic igneous rock intermediate in crystal size between fine-grained basalt and coarse-grained gabbro. Specifically, it's the Mashonaland Diabase, quarried in Zimbabwe's Mashonaland East Province. This stone formed when magma worked its way upward into older rock to form tabular structures called sills and dikes. These were created during an episode of volcanic activity and continental collision about 1.8 Ga, during the Paleoproterozoic era. The Mashonaland Diabase, also known by such hypercaffeinated trade names as "Nero Assuluto" and "Absolute Black Zimbabwe," owes its somber tone to the deep shades of its primary minerals: amphibole, augite, and normally lighter plagioclase here darkened by inclusions of magnetite.

5.16 Kesner Building

5 N. Wabash Avenue
Completed in 1910
Architectural firm: Jenney, Mundie & Jensen
Foundation: Rock caissons
Geologic feature: Larvikite Monzonite

This Jewelers' Row site features one of the showiest and most curiosity-provoking cladding stones found in the city. It's on display here in the framing of the building's main doorway, at the northern end of the façade, and it probably postdates the building's construction. At first glance this arresting stone could be mistaken for a "black granite"—a gabbro, diabase, or anorthosite. But embedded in its matrix of dark pyroxene crystals there is a lighter one, perthite, that flashes and glitters iridescently in the sun. Perthite is actually two types of feldspar that have grown together; the light is reflected at the boundary surfaces between them. This amazing effect goes by the name of labradorescence, or, less ponderously, schiller.

The highly polished stone here is the darker "Emerald Pearl" variety of Larvikite Monzonite, a type of intrusive igneous rock that differs from granite by either having very little quartz, or none at all. It draws its name from its quarrying region near Larvik, Norway, southwest of Oslo. Now widely popular as a decorative stone, Larvikite Monzonite originally formed underground as part of a swarm of separate plutons early in the Permian period.

5.17 Stevens Building

16 N. Wabash Avenue
Completed in 1913
Architectural firm: D. H. Burnham & Company
Foundation: Hardpan caissons
Geologic feature: Conway Granite

When one thinks of granite, the color green doesn't usually spring to mind. And indeed most rocks of that type have a gray, off-white, pink, or reddish cast. But on the trim around this building's main Wabash Street entrance you'll find stone marketed as "Redstone Green," a rather offbeat version of the Conway Granite. Elsewhere on view in the city there's the green variant of Massachusetts' Cape Ann Granite. Here it's another type, quarried in days past on Rattlesnake Mountain in Redstone, an unincorporated community within the town of Conway, New Hampshire. (Why *Red*stone? Because the more common "Conway Pink" variety was also produced there.)

The green variant of the Conway Granite on the Stevens Building is typical of its appearance when polished. It most definitely isn't imbued with the rich, dark green of a serpentinite. Rather, it's an essay in that color at its most equivocal—a vague and hazy sort of greenishness whose existence can be completely confirmed only when you get up close and discern the olive-tinted quartz crystals tucked among the gray orthoclase feldspar and black biotite and hornblende. Isotopic dating indicates that the Conway Granite came into being as part of the massive White Mountain Batholith in the Jurassic period. It was a time when New Hampshire became the focus of massive magmatic upwelling and explosive volcanic activity. The crust stretched, thinned, and fractured as what is now northwestern Africa separated from the Eastern Seaboard, during the breakup of the supercontinent Pangaea and the birth of the North Atlantic Ocean.

5.18   Shops Building

17 N. Wabash Avenue
Completed in 1875, substantially renovated in 1912
Architects: Unknown (1875), Alfred S. Alschuler (1912)
Foundation: Shallow and isolated spread footings
Geologic feature: South Kawishiwi Troctolite

Decked out in standard Jewelers' Row fashion with a main entrance framed in polished stone, the Shops Building is geologically notable for its display of the decidedly showy South Kawishiwi Troctolite, a flamboyantly beautiful rock type named for a river in its North Woods quarrying region. Still on the market today, it has been extracted since the early 1900s from a remote location in Superior National Forest, southeast of Ely, Minnesota. This particular variety of the South Kawishiwi was marketed in its day as "Minnesota Black Granite." Mesoproterozoic in age and a product of the great Midcontinent Rift event that almost tore our continent apart, it's a coarse-grained mafic igneous rock much more closely related to gabbro than granite. It is mostly composed of labradorescent plagioclase feldspar, with minor amounts of augite and olivine. Glittering in sunbeams not fully filtered by the El tracks above, the stone's large plagioclase crystals seem to suggest the cut gems beckoning to buyers from nearby store windows.

5.19   Pittsfield Building

55 E. Washington Street
Completed in 1927
Architectural firm: Graham, Anderson, Probst & White
Foundation: Rock caissons
Geologic features: Mellen Gabbro, Rosso Ammonitico Veronese Limestone, Holston Limestone, Botticino Limestone, Sylacauga Marble, Tinos Ophicalcite, Northwestern Terra Cotta, Bronze, Brass

This skyscraper is Gothic Revival in style with a slight Art Deco accent. It's a member of the select club of buildings that have been, at some point and however temporarily, Chicago's tallest. It's also one of the Windy City's best architectural showplaces of geology and paleontology. In fact, this place is a veritable museum of beautiful stone types. While most of its exterior cladding takes the form of light-gray Northwestern Terra Cotta, the ground level is adorned with a handsome blend of bronze trim and Mellen Gabbro, a brand of "black granite" that was a favorite of Chicago architects in the 1920s. Quarried in northern Wisconsin and Mesoproterozoic in age, the Mellen is here represented by its "Rosetta Black" variety, which possesses a dark matrix of plagioclase and ferromagnesian minerals with a smattering of labradorescent crystals that shimmer in a way reminiscent of Larvikite Monzonite. The Mellen formed underground during the great Midcontinent Rift event at about 1.1 Ga, when the continental crust of what is now the Lake Superior region began to split apart. This rifting ceased before what is now Canada could break away from the North-Central United States. Had it not, Chicago would probably now belong to some land mass other than North America.

The decorative panels of the Pittsfield's entrances couldn't be more different in appearance. They're made of a celebrated limestone, the Rosso Ammonitico Veronese. When spoken by Italian geologists this sesquipedalian name rolls off the tongue sweetly; but even they tend to abbreviate it into "RAV." At first glance this stone, which is best known in the building trade as "Rosso Verona Marble," looks unsettlingly like gnocchi floating in a bowlful of tomato soup. Amid the pale nodules the persistent searcher will find, especially in the smaller doorways facing Washington Street, some superb and readily recognizable fossils—the large coiled shells of ammonites, the prehistoric relatives of modern squid, octopi, and cuttlefish. These marine invertebrates were plentiful in the warm waters of the Tethys Ocean, which covered what is now the Mediterranean area and southern Asia during the Jurassic period. Their remains and the rock you see here were deposited on a drowned, basin-bordered plateau. In architectural use since ancient Roman times, this stone is quarried in the area to the north and northwest of Verona, Italy.

The Pittsfield is one of the Loop buildings that has its interior decorative stone and metal well documented. The best place to start exploring it is at the ground level of the lovely five-floor atrium accessible from both the Wabash and Washington entrances. Along with all the brass ornament, distinguishable from the exterior's bronze by its lighter, more yellow-golden tint, there are flooring and basal wall trim in complementing tan and deep-pink versions of the Ordovician-period Holston Limestone, otherwise known as "Tennessee Marble." Above it the main wall cladding is another Italian import of Jurassic heritage: the beige,

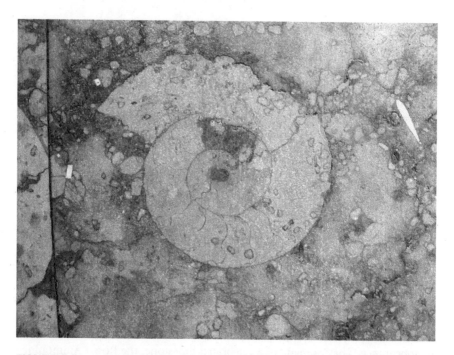

FIGURE 5.5.   Jurassic Chicago. A beautifully preserved ammonite shell, set on a background of red with lumpy pink nodules, gives visitors passing through a Pittsfield Building entrance a glimpse into the abundant marine life of the great Tethys Ocean in the age of the dinosaurs. This is the famous "RAV": the Rosso Ammonitico Veronese Limestone quarried in northern Italy.

mottled, and sometimes fossiliferous Botticino Limestone. Like the Holston, the Botticino is a common sight in the lobbies and hallways of Chicago's finer buildings. At this point it should come as no surprise that it, too, is a stone the building trades compulsively call a marble. After all, in their parlance just about any softer rock that takes a good polish qualifies for that category.

The visitor willing to venture on to either the Pittsfield's upper levels or to the lower arcade accessible by staircase from the atrium will find still other notable stone selections. In addition to more pink Holston Limestone flooring it's easy to spot the gleaming white Sylacauga Marble—yes, at last a true marble—and the deep-green Tinos Ophicalcite used for the wall bases on the floors above the atrium. The Sylacauga, named for the Alabama community where it was produced, is thought to be of early Paleozoic, and probably Cambrian or Ordovician, age. It's composed of interlocking calcite crystals and is often of exceptional fine-grained purity. As such, it is second in quality to no competitor, including the much more famous Carrara Marble of northern Italy.

On the other hand, the green stone here from the Greek isle of Tinos is our final example of a marble-that-isn't. Instead it's an ophicalcite and as such shares the same coloration, darkling aspect, and exotic origin as the Vermont Serpentinite on view across the way at the Cultural Center. The difference between the two can be hard to make out, but the discerning eye will catch that ophicalcite is more thoroughly brecciated—with a texture of angular chunks of dark rock spaced within a more intricate webbing of white veins. As ephemeral surface creatures, it's hard for us to imagine what this stone has already gone through. Recent geologic studies suggest that it began to form in the Mesozoic era, in the Earth's interior above a Tethys Ocean subduction zone, where crust of one tectonic plate was sinking under the crust of another. In this environment both mafic igneous rocks and ocean-floor sediments were present. Instead of being swept farther downward, this jumble of rock types was eventually thrust up by an approaching continental fragment and scrunched into its current position in the Cyclades island chain southeast of Athens. While it experienced various phases of transformation after its parent materials formed, it's generally considered to be Cretaceous in age.

5.20 Garland Building

111 N. Wabash Avenue
Completed in 1922
Architect: Christian A. Eckstorm
Foundation: Rock caissons
Geologic feature: Milbank Area Granite

All scientific systems of classification establish boundaries between the things they attempt to describe and delimit. But sometimes those boundaries can be frustratingly fuzzy. This Jewelers' Row site is an excellent example of that fact.

The point of interest here is the cladding that frames the main entrance along Wabash. This stone is, to my eye at least, the most attractive of the various varieties of the ancient, Neoarchean-age Milbank Area Granite that adorns many Chicagoland buildings. Marketed as "Agate," it has the characteristic ruddy-to-muddy-red coloration of the Milbank group, which is quarried in communities straddling the Minnesota-South Dakota border. The Garland Building's stone has a good deal of black biotite in addition to the quartz, orthoclase, and microcline that provide its somewhat lighter tones. But it also often has a foliated aspect with wavy bands, very visible here, that suggest it's a metamorphic gneiss, or on its way to being one, rather than a purely igneous granite. And indeed some authorities have noted this and given it the modifier "gneissic." Here we're in the borderlands between one fundamental grouping and another. Perhaps we should be reassured and even exhilarated by the fact that nature stubbornly insists on presenting the occasional transcendent ambiguity.

5.21   Marshall Field & Company Store

111 N. State Street
Current name: Macy's on State
Completed in separate sections in 1892, 1902, 1906, 1907, and 1914
Architectural firms: D. H. Burnham (1892), D. H. Burnham & Company
   (1902, 1906, 1907), Graham, Burnham & Company (1914)
Foundations: Different types, see the discussion below
Geologic features: North Jay Granite, Concord Granite, Terra-Cotta,
   Bronze, Favrile Glass

I retain this site's original name to clarify its complicated history and its relationship to site 5.22, the Marshall Field & Company Annex. The massive store complex described here occupies an entire block bounded by Wabash Avenue and State, Randolph, and Washington Streets. In addition to being a showplace for New England granites, it's a wonderful demonstration of the evolution of building foundations in Chicago, from those on shallow footings to caissons that reach all the way to the bedrock far below. In all there are five interconnected sections, three of which face Wabash and two that face State. All are twelve stories except for the oldest surviving section, which is nine. Various sources cite such a surrealistic assortment of stone types for the exterior of the extant building that, for the geological researcher, this site is the architectural equivalent of a greased pig. Previously held understandings squirm all to easily out of one's grasp. Part of the problem is that architectural historians, presumably very attentive to accuracy in other matters, often use rock terms with complete artistic abandon. So the following section-by-section descriptions represent my best detective work to date.

The 1892 Section

This Italian Renaissance–inspired building is situated on the northwest corner of Wabash and Washington. The oldest surviving portion of the complex, it's sometimes still confusingly called "the Annex," a name now better applied to the store addition across Washington Street. Here the exterior walls are actually load-bearing; at the time of construction the concept of iron- or steel-frame construction was still in its infancy. The foundation is a shallow one on isolated spread footings composed of beam and rail grillages—a common design before the advent of the deep-reaching caisson.

The exterior is stone on the lowest three stories and buff terra-cotta above. At more than a pace or two away the rusticated ashlar at ground level looks like a coarsely crystalline marble, but it doesn't react in the least to dilute hydrochloric acid, the geologist's classic field test for rocks containing calcite. But it does have a few telltale, glittery muscovite flakes that reveal its igneous origin. This is the Devonian-age North Jay Granite of Maine, albeit in a rather deteriorated state, with most of its exposed biotite and plagioclase-feldspar crystals weathered out and the muscovite and white potassium-feldspar crystals still intact. This tally of mineral victims and survivors is in consonance with a fundamental geologic concept known as Bowen's reaction series, which details both the temperatures that various compounds crystallize at in a cooling magma, and also how long they're chemically stable in the much less infernal setting of the Earth's surface. The minerals mostly missing here formed at higher temperatures but are less enduring, and have broken down at the surface of the stone. This deterioration of the North Jay provides us with a good lesson in how even granite can pit, spall, and scale the way less resilient rock types do. And it brings to mind another wall, one excavated in Pompeii, where some ancient hand had scrawled NIHIL DURARE POTEST IN TEMPORE PERPETVO (nothing is capable of enduring forever). For both living and nonliving things, this is the grand and brutal lesson of Earth's geologic and biologic cycles.

The 1902 Section

Standing diagonally across from the 1892 edifice, at the southeast corner of State and Randolph, this portion features a deep foundation that signals the next evolutionary step in this city's civil engineering: caissons dug not all the way to the bedrock but to the stable layer of hardpan that was found about 85 feet down. It can be difficult on the outside to tell where this section begins or ends, because it was ultimately merged with the 1907 section south of it to create one harmonious Beaux Arts façade with Chicago windows. The stone at ground level, and apparently at least part of the way to the top, too, is the light-gray, salt-and pepper Concord Granite, but in one sense it's quite distinct from other stone of this name on display in Chicago. For it was quarried not at the main source in Concord, New Hampshire, but some forty miles away, at the Webb Quarry in Marlborough, a small town in the southwestern portion of that state. This is a coolly classy selection that can be distinguished by its minutely fine-grained texture and spar-kling plates of silvery muscovite mica. It, too, dates from the Devonian period.

The 1906 Section

This is the middle portion of the Wabash-facing side. Constructed just four years after the preceding, it features the store's first set of caissons dug all the way to the bedrock. The stone seems to be Concord Granite, too, or something very similar. It shows signs of spalling here and there.

The 1907 Section

In this southern part of the State Street side, a caisson foundation was also used, and reportedly its shafts were dug from 90 to 110 feet down to reach the underly-ing Silurian dolostone. On the building's exterior the cladding once again is Con-cord Granite, as is the case, apparently, for the four huge Ionic columns flanking the main entrance, even though various architectural historians have erroneously listed them as marble or limestone. Inside, hovering in polychromatic glory high above the ground floor, is a different geologically derived wonder: a magnificent rectangular dome ornamented with a mosaic shimmering with 1.6 million pieces of Tiffany Favrile Glass.

Also, I'd be remiss to pass over the Loop's most beloved points of reference, the pair of matching and ornately designed clocks mounted 17 feet above grade here at the State-and-Washington corner and at State and Randolph on the 1902 section. The green patina of these giant timepieces confirms that they're encased in that venerable alloy of copper and tin, bronze. And, at about 7.5 tons each, they have a lot of it.

The 1914 Section

Here at this constructional saga's end, at the southwestern corner of Wabash and Randolph, we once again have a foundation of rock caissons and what appears to be a continuation of the Concord cladding. However, for this whole block caveats abound: the Concord and North Jay, especially in certain types of finish and when covered with urban grime, can be devilishly tricky to identify with cer-tainty. On top of that, one reliable source also list's Maine's Devonian Hallowell Granite for this building. It may have been used as exterior cladding on the newer sections above the seventh floor, but it may be present as well somewhere within easier reach. As the geologist Robert Folk wrote in his classic work on sedimen-tary petrology, "None of the statements herein are to be regarded as final; many ideas held valid as recently as two years ago are now known to be false. Such is the penalty of research."

5.22   Marshall Field & Company Annex

25 E. Washington Street
Completed in 1914
Architectural firm: Graham, Burnham & Company
Foundation: Rock caissons
Geologic features: Cape Ann Granite, Salem Limestone

Separated from the main, block-filling Marshall Field's store by the width of Washington Street, this building is nevertheless part of the general construction history of the whole. But in some ways it's quite distinct. It stands twenty stories tall to the other sections' nine to twelve, and, with the exception of the pilasters that rise from the first to the second floor, it's clad not in granite but in Chicago's most prevalent architectural rock type, the Salem Limestone. But granite there is, albeit of a different sort, on those pilasters. And like that of the main store building across the street, it's a variety that requires some careful investigation to avoid misidentification.

The New England geologist T. Nelson Dale, writing just nine years after this edifice was completed, states that the North Jay Granite was the stone chosen for Chicago's "Field Annex Building." That certainly sounds like a clear-cut reference to this site, but then I discovered that in earlier decades the name "Annex" referred instead to the 1892 section of the main store complex across the street. This is just more evidence that practically everything about the architectural geology of Marshall Field's is more complicated than it first seems.

And indeed I found on close examination that the exterior's granite cladding here just can't be the North Jay. Gray it is, but fine-textured it isn't. Fortunately, when I then combed through the 1960s Coldspring list of Chicagoland granites, I found its anonymous compiler had identified the stone here as "Rockport Gray." That's a trade name for the Cape Ann Granite, a handsome blend of light-toned orthoclase, smoky quartz, and black hornblende. It was quarried along the rugged Atlantic coast, on Massachusetts' North Shore. And careful comparison with other documented Cape Ann Granite sites in Chicago shows the stone here is indeed a perfect fit for it.

Unlike the granites adorning the main store, the Cape Ann is Silurian in age, and thus one geologic period older than they are. While it was taken from what now forms the eastern fringe of North America's crust, it began altogether elsewhere, as a pluton or batholith in the depths of the microcontinent Avalonia.

Only later, in the Devonian, did that wandering land mass and its contents, including the Cape Ann Granite, merge with what is now the US mainland.

5.23   Reliance Building

32 N. State Street
Current name: Hotel Burnham
Foundation: Shallow and isolated spread footings of the grillage type
Completed in 1891 (first and second stories); 1895 (upper stories)
Architects: John Wellborn Root of Burnham & Root (1891), Charles B.
    Atwood of D. H. Burnham & Company (1895)
Geologic features: Northwestern Terra Cotta, Boston Valley Terra Cotta,
    Aberdeenshire Granite (now apparently replaced by another granite)

It may be hailed as a forerunner of modernist skyscrapers, but the Reliance is no mere glass box. Its magnificent array of Chicago windows sits in a grid of white terra-cotta best appreciated from across the street. As the noted architectural historian Carl Condit put it,

> The building is a triumph of the structuralist and functionalist approach of the Chicago school. In its grace and airiness, in the purity and exactitude of its proportions and details, in the brilliant perfection of its transparent elevations, it stands today as an exciting exhibition of the potential kinesthetic expressiveness of the structural art.

Originally the contrasting stone cladding on the lowest two floors was, according to contemporary accounts in the *Architectural Record* and the *American Architect*, "Scotch Granite." Since it was a red variety, it may well have been the type of Aberdeenshire Granite quarried in Peterhead, Scotland. It was a popular choice in America at the time. During a major renovation in the 1990s only one surviving fragment of it was discovered, and a similar red granite, the identity of which I haven't been able to ascertain from the restoration architects, was chosen for the recladding. Whatever it really is, it's a striking stone indeed, and it does closely resemble the Aberdeenshire.

The more than 14,000 sections of enameled terra-cotta that stand above the granite were originally produced by Chicago's Northwestern works, but during the restoration a significant portion of them were removed. Some were refurbished and reset; others had to be replaced by exact replicas crafted by the Boston Valley works of western New York. So we can once again see the Reliance

Building very much as it appeared when it impressed a correspondent writing in an 1895 issue of the trade journal the *Brickbuilder*:

> A notable feature of the "Reliance" is the terra-cotta, which is all glazed, and, being cream- white in color, presents the appearance of a porcelain building. A semi-annual water bath will make this building always new. The Northwestern Terra-Cotta Co. are to be congratulated on the uniformity of results obtained in the color and the glaze.

5.24 Carson Pirie Scott & Company Store

1 S. State Street
Original name: Schlesinger & Mayer Store
Current name: Sullivan Center
Completed in 1899 (original section), additions in 1904, 1906, and 1961
Architects: Louis Sullivan (1899 and 1904), D. H. Burnham & Company (1906), Holabird & Root (1961)
Foundations: Different types, see the discussion below
Geologic features: Cast Iron, Northwestern Terra Cotta

Like its competitor, the Marshall Field complex just two blocks to the north, the Carson Pirie Scott Store (usually shortened to "Carson's") represents a progression of construction projects, the two most substantial being those of 1899 and 1904. The final two simply added floor space southward on State Street with no significant stylistic changes. The 1899 section faced Madison Street a little east of State. Its foundation consists of a series of wooden piles driven 50 feet into the substrate. By 1904, however, Chicago engineering had squarely entered the age of caissons, and those used for the second section of the store were dug all the way to bedrock.

But unlike the Field complex, the Carson's store is no open-air museum of American stone types, even though architect Louis Sullivan reportedly first planned to use a white variety of Georgia's Murphy Marble for its exterior cladding. Instead, the building's realized design owes its powerful ornamental impact to two other elements, its enameled white Northwestern Terra Cotta Company cladding and Sullivan's endlessly fascinating iterations. As noted in the Commission on Chicago Landmarks report on this building, these designs were not simply the product of one brilliant man, but of a veritable assembly line of artistic genius. Sullivan's initial concepts and overarching artistic vision were enhanced and developed by draftsman George Elmslie, destined to be a great architect in his own right; then his sketches were transformed into three-dimensional moldings by the Norwegian-born model-maker Kristian Schneider. These in turn were used to cast the finished product by the Winslow Brothers Company.

In particular, the dazzling detail and artistry of the famous cylindrical street-corner entrance has to be seen, from both outside and the vestibule within, to be believed. Obviously these flights of highly ordered fancy were cast into metal, but when I ask my tour participants to identify it, those who don't already know the building's history invariably guess that it's bronze. After all, there's that greenish-brownish-reddish luster so suggestive of that alloy in its various weathered forms. But in reasoning so, they're doing precisely what the architect wanted them to do. The medium Sullivan employed here is actually the less expensive cast iron artfully painted to simulate the classier medium. And the process of coating it was painstaking. First, to protect this metal that is so prone to oxidation and corrosion, a sealant of *asphaltum* was brushed on. (This is the same sort of tarry organic substance found occurring naturally in our region's reefal Silurian dolostone.) Then a layer of bright red paint was added, and over that a variably opaque outer lamina of green, which in places artfully reveals the ruddy tints of the lower layer.

5.25   Heyworth Building

29 E. Madison Street
Completed in 1905
Architectural firm: D. H. Burnham & Company
Foundation: Caissons (type not specified)
Geologic feature: Deer Isle Granite

The exterior trim of this building's main entrance is worth a quick look. The "black granite" here has not been identified, but the lighter stone within it is the handsome Deer Isle Granite, quarried on Crotch Island near the Atlantic seacoast town of Stonington, Maine. This rock type formed as one of the Devonian intrusions that were triggered by the collision of the microcontinent Avalonia with the mainland of Laurentia, the precursor of modern North America. The coarse-grained crystals visible here include beige orthoclase and microcline, dark-gray quartz, white oligoclase, and black biotite.

THE LOOP
Southeastern Quadrant

6.1 Monroe Building

104 S. Michigan Avenue
Completed in 1912
Architectural firm: Holabird & Roche
Foundation: Rock caissons
Geologic features: Salida Granite, Northwestern Terra Cotta, Rookwood
    Ceramic Tile

Located at the head of the street from which it takes its name, the gabled Monroe Building is a blend of Romanesque and Italian Gothic styles. It has a special claim to geologic notoriety; the stone that adorns its lowest two stories is what is probably the least frequently encountered granite in Chicago architecture.

This rarity is the Salida Granite, here represented by its "Salida Pink" variety. While it was processed for use and shipment in the Colorado town of that name (which, by the way, is pronounced "sah-LIE-duh" by its inhabitants), the exact quarrying location of the Salida Pink type is still open to question, despite hours of researching on my part and that of Earle Kittleman, of the Salida Museum Association. The main reasons for this ambiguity, we discovered, are that the production of this variety ceased half a century ago, and that other varieties from at least three different quarries in the region were marketed under the Salida brand name. Still, the two most likely quarries for the Monroe stone lie along the Ute Trail, about 6 miles north of town, and some 25 miles to the southeast, near Texas Creek in neighboring

Fremont County. Currently my instincts incline to the latter. But wherever it came from originally, the Salida Granite on view here is a beautiful, light-pink, coarse-grained rock dating to either the Paleoproterozoic or Mesoproterozoic. Decorative stone this striking deserves to be used and seen more than it has been in Chicago.

Above the building's Salida base the cladding is terra-cotta produced locally by the Northwestern works. The style selected here was marketed as "Standard Granite" because it features varicolored spotting on a light background that simulates the matrix of mineral crystals found in the real igneous rock. This material had two separate trips through the kiln—first, to form the uncoated *bisque*, the fired clay serving as the base for the coating, and again, to fix the enamel itself.

Much more accessible to close examination is the matchless display of glazed ceramic tile, also referred to as *faience*, in the building's lobby. This is an interior space that, for all of its stiff competition, must be one of the most gorgeous in the entire city. The ceramic tile was produced by the famous Rookwood works in Cincinnati. Remarkably, much of it here had been covered up during a midcentury remodeling project. However, in 2009 and 2010 a full renovation once again revealed the original 1912 Rookwood decoration and supplemented it with new tiles from the same company. These were painstakingly created to duplicate the ageing coloration and craze lines (some indelibly stained with grime) of the originals. While its glaze is the product of the subtle interactions of a host of chemical compounds skillfully manipulated by Rookwood chemists and ceramists, the clay at its base is one of Earth's commonest—and most amazingly useful—materials available to the artist and architect. It's the end state of the disintegration of rock, a residue of tiny, cohesive, and platelike mineral particles each no larger than one ten-thousandth of an inch. But in geology end states are always elusive: in the grand cycle of destruction and creation clay can later form a new stratum of rock or, if shaped by the human hand and mind, some other thing of great utility and beauty.

6.2   Illinois Athletic Club

112 S. Michigan Avenue
Completed in 1908
Current name: School of the Art Institute of Chicago
Architectural firm: Barnett, Haynes & Barnett
Foundation: Hardpan caissons
Geologic features: Concord Granite, St. Cloud Area Granite, Salem
   Limestone

The exterior of most of this building is clad in that great purveyor of tranquil and uniform grayness, the Salem Limestone or "Bedford Stone" of Hoosier origin and Mississippian age. It's an excellent choice for the rusticated ground-floor façade, especially because it hasn't been allowed to come into contact with the sidewalk and its deadly brew of saline meltwater. This danger has been avoided, at least somewhat, by the provision of a damp course of the less porous Concord Granite, which from a distance is almost indistinguishable from the Salem. But it does yield up its differences to the urban geologist wielding a hand lens.

The Concord, quarried in the New Hampshire city of that name, is another one of those New England granites that owes its existence to the intense tectonic activity and upwelling of magma triggered by the merging of Avalonia and other wandering crustal fragments with ancestral North America in the Devonian period. It's fine- to medium-grained and tends to be at least a little porphyritic. This means that some of the crystals are larger than the rest, as indeed the white oligoclase feldspars often are here, when compared with the surrounding blend of gray microcline and quartz, silvery muscovite, and black biotite. (The muscovite crystals, flashing where exposed on the surface, make a particularly good Concord identifier.) On this building the granite, though generally more resistant than the Salem, hasn't remained completely impervious to the attack of deicers. It's exfoliating in places where pavement meets stone.

The dedicated rockhound will find one other item of interest: a small sampling of a coarser-grained stone on the lower surfaces facing the two doors that flank the main entrance. This is the relatively rare "Diamond Gray" variety of the otherwise very extensively used St. Cloud Area Granite, which is quarried in a wide assortment of textures, tints, and tradenames from the Paleoproterozoic batholith that underlies central Minnesota.

6.3   Peoples Gas Building

122 S. Michigan Avenue
Completed in 1911
Architectural firm: D. H. Burnham & Company
Foundation: Hardpan caissons
Geologic features: Cape Ann Granite, Milford Granite, Northwestern
     Terra Cotta, Bronze

The most immediately eye-catching design feature of this striking building is its collection of eighteen stout cylindrical shafts, each weighing 26.6 tons, with Ionic capitals. These stand at attention at the base of the Michigan and Adams façades like a drill-team squad of giants. At 26 feet from top to bottom, they're only a little more than half the height of the columns of the same architectural order that front the Marshall Field's Store on State Street. But their visual impact is ever so much greater. Each was fabricated of the "Rockport Gray" variety of Cape Ann Granite quarried on the windswept promontory of that name north of Boston. This is the optimal place in the city to examine this medium- to coarse-grained Silurian rock type in all its polished glory. Technically classified as an alkali-feldspar granite, the Cape Ann is a striking medium- to coarse-grained rock with a rich patterning of black hornblende set against dark glassy gray quartz and light bluish-gray orthoclase and microcline, which are the alkali feldspars that here are found to the exclusion of other feldspar types.

But the Cape Ann is not the only Massachusetts granite present. On the first two floors, recognized most easily in the cladding behind the columns, is the pale-pink Milford. Dating to the Neoproterozoic, it's definitely the older of the two rock types, though like the Cape Ann it comes from the Bay State's Avalonian terrane, which began as a wandering microcontinent that collided with North America in the Devonian period. Unlike the Cape Ann, however, it contains both alkali and plagioclase feldspars—the former pink, the latter white to a pale greenish yellow. It is also distinctively marked with large crystals of black biotite set against the generally pink background. All in all, it's one of the most beautiful of the New England granites.

If the Peoples Gas Building is a showplace for real granite, it's also a chastening lesson in brilliantly deceptive faux granite. For the cladding on the Michigan and Adams corner-pier sections containing the entrances and pilasters looks to all intents and purposes just like the Milford, replete with the same color scheme and varying matrix of light and dark crystals. But in this case it's time to pull out the hand lens and examine whatever bolt holes or chipped-off spots one can find: the cladding interior is not made of more crystals, as it should be. Instead, it's a whitish *bisque*, the fired clay that serves as the substrate for the glazing in terra-cotta. This wonderful stand-in for real stone, so cleverly created by the artisans of Chicago's Northwestern works, also constitutes the fancy-patterned cladding from the fourth floor to the top of the structure, and even the upper row of columns.

Also of note on the ground story's exterior is the liberal use of bronze trim and framing. A total of 30 tons of this time-honored alloy of copper and tin was installed here. This certainly sounds like a lot, but it's not much more than the weight of just one of the Cape Ann Granite columns. The metal has been allowed to take on the distinctive bright-green patina that is the common sign of weathered bronze. This weathering product is composed of copper salts—the result of a kind of corrosion, but one very much to be desired, because it forms a self-sealing, protective coat that retards still more chemical change.

6.4 Art Institute of Chicago

111 S. Michigan Avenue
Completed in 1893, various subsequent additions
Architectural firm: Shepley, Rutan & Coolidge for the original section
Foundation: Shallow and isolated spread footings
Geologic features: Salem Limestone, Concord Granite, Bronze

No other building exterior in the city offers the urban geologist greater opportunity to examine, at nose-flattening hand-lens range, America's most common architectural rock type, popularly known as the "Bedford" or "Indiana Limestone" but better identified by its geologic tag of Salem Limestone. My favorite spot to do so is at the balustrade and building exterior fronting the Art Institute's South Garden. There a grove of sculptured hawthorns provides shade and relief from the clumps of tourists who drift like herds of wildebeest on the sidewalk, and from the eardrum-shattering creativity of the street musicians by the main entrance. The stone as you see it now is much cleaner than it was in the age of wholesale bituminous-coal burning. Within a year of its opening, the *American Architect* noted, the Art Institute had "lost all chance of taking on its time-growing richness of color, because of the coating of soot and smoke." Modern Chicagoans annoyed by the momentary whiff of sewer gas or the exhaust fumes

ART INSTITUTE, MICHIGAN BOULEVARD, BETWEEN MONROE AND ADAMS STREETS.

FIGURE 6.1. Originally published in *Rand, McNally & Co.'s Bird's-Eye Views and Guide to Chicago*, this 1893 rendering of the newly completed Art Institute reveals its proximity to what was then the lakeshore. Note the pedestrian bridge over the water-fronting Illinois Central Railroad tracks that provides access to a docked excursion steamer. This vessel was used to transport visitors staying downtown to the Columbian Exposition, situated 6.5 miles to the south. If you look closely you'll also spot Edward Kemeys's bronze lions, already standing guard duty in front of the museum.

of a passing bus should keep in mind that their city's atmosphere was once considerably more polluted than they can now imagine.

In its natural outcropping form the Salem can be seen not only in the great southern Indiana quarrying district centered on the town of Bedford, but also in three adjoining states, including in Illinois, along the Mississippi River bluff from just below East St. Louis to Prairie du Rocher, and also in a separate band north of Cairo. This type of limestone, biocalcarenite, is composed of an unimaginably abundant supply of tiny whole fossils and hard-part fragments from larger marine invertebrates, all cemented convincingly into rock with the mineral calcite. It may be difficult to imagine in this leaf-dappled courtyard, but the stone that here has been so shaped to human ends originated in a tropical sea, in the Mississippian portion of the Carboniferous period, during the global sea-level highstand known as the Kaskaskia Sequence. The Mississippian sea varied in extent as its level rose and fell, but at its maximum covered much of the Midwest. The Salem can almost be mistaken for sandstone, especially when seen in the rough. It owes its special granular, porous texture to the environment that spawned it: shallow carbonate shoals, tidal channels, and lagoons teeming with a diverse fauna. In those ancient days, what is now Indiana and Illinois lay close to the equator. North America, then combined with sections of other continents, was well on its way to colliding with the Gondwana supercontinent to produce the even larger land mass of Pangaea.

All the automotive and pedestrian commotion notwithstanding, a stop just outside the main entrance itself is a must. Here you can take in to best advantage the Art Institute's grand neoclassical façade, with its rusticated Salem ashlar below and the corresponding dressed-face courses above. This spot also features two other points of geologic interest, both of which involve one of the Windy City's most beloved icons, the pair of bronze lions sculpted in 1893 by Edward Kemeys and unveiled on-site the following year. Cast by Chicago's own American Bronze Founding Company, each of these famous quadrupeds weighs more than 2 tons. The *American Architect*'s Chicago correspondent also wrote a fulsome description of the statues as they were being made:

> The French superintendent of the foundry, Jules Bercham, is sparing neither time, money nor skill to make the casting as successful as possible. An amusing little fact connected with this part of the work is that the tails have been made nearly twice as thick as the bodies, for the tails of such lions, as everyone knows, are a favorite perch for the terrible and omnipresent small boy, and as his habits cannot be cured, preparations are thus made to make them as harmless as possible.

Over the years the lions have been allowed to naturally develop a striking patina. This coating, often described as "copper oxide," is in fact the result of a surprisingly complicated daisy chain of chemical reactions that merely starts with oxide compounds, which by themselves are usually pinkish or black. So what's on view here is the stable end point, a medley of copper salts—some combination of copper sulfate and two different forms of copper carbonate. It is they that produce the classic green tint.

The pedestals on which the lions stand match the color of the main building's Salem Limestone, but when inspected at close range turn out to be a chaste gray granite instead—a very good choice for this deicer-splattered, hustled-and-bustled setting. (Some of the badly weathered carved detail in the Salem nearby shows how challenging the local environment is.) A whole bevy of different granites have been documented for the Art Institute without their specific locations cited. From that tally, I've surmised that this one is probably the upper-Devonian Concord Granite of New Hampshire. I base this on its color, fine- to medium-grained and rather porphyritic texture, the presence of larger white oligoclase or albite feldspar crystals, and its specks of muscovite mica that sparkle in the sunlight.

6.5   Borg-Warner Building

200 S. Michigan Avenue
Completed in 1958
Architects: William Lescaze and A. Epstein
Foundation: Rock caissons
Geologic feature: Hägghult Diabase

A vertically stacked glass-and-metal shoebox, the 20-story Borg-Warner Building might seem sterile ground for the rockhound, and a much better subject for meteorologists because of the way it reflects the passing clouds so nicely. But decorative stone has had its way of insinuating itself into and sometimes onto even the most anonymous of International Style skyscrapers. Here the black piers that punctuate the large ground-floor windows are a geologic treasure, because their cladding is Hägghult Diabase, otherwise known as "Black Swede" or "Swedish Black Granite." This dark and elegant rock type owes its name to the locality where it's quarried, in Skåne, Sweden's southernmost county.

This stone takes a very high polish: on the Adams Street side it mirrors passing traffic, pedestrians, and the colonnade of the Peoples Gas Building across the way to a remarkable degree. Diabase, being intermediate in grain size between basalt and gabbro, is characteristically formed in bodies of magma—usually dikes or sills—that solidify underground, but not far from the Earth's surface. The Hägghult has been dated to 1.6–0.9 Ga, which places it in the lower Mesoproterozoic to lower Neoproterozoic eras. Over half of its mineral content consists of types of plagioclase feldspar. Its other constituents are amphibole, biotite, chlorite, and serpentine. Some of the crystals produce a nice play of colors in the sunlight.

The Borg-Warner Building is notable from an engineering standpoint, as well. Its rock caissons were reportedly the first in Chicago to be machine-drilled rather than dug out by hand.

6.6   Symphony Center

220 S. Michigan Avenue
Completed in 1904
Former names: Theodore Thomas Orchestra Hall, Orchestra Hall
Architectural firm: D. H. Burnham & Company
Foundation: Wooden piles
Geologic feature: Salem Limestone

The Georgian façade of this home of the planet's greatest symphony orchestra features the tried-and-true Salem Limestone set in a different key. Here it isn't the main cladding material, as it is across the boulevard at the Art Institute and on other landmarks nearby. Instead, it plays a highly effective second fiddle as the harmonizing decorative details of base, quoins, frieze, window dentil cornices, balustrades, and framing. This all serves as an effective leitmotif embedded within the main theme of handsome red brick, which, unfortunately, is of unknown origin. While it might have irked contemporary fans of Berlioz, Dvořák, Tchaikovsky, or Verdi, the Teutonocentric frieze bears the names of the four composers held in highest esteem at the time. Note how cleanly engraved in the Salem those names remain after twelve decades of exposure to the caustic urban air.

6.7   Railway Exchange Building

224 S. Michigan Avenue
Completed in 1904
Currently known by its street address
Other former names: Santa Fe Building, Santa Fe Center
Architectural firm: D. H. Burnham & Company
Foundation: Hardpan caissons
Geologic feature: Northwestern Terra Cotta

Contrasting mightily with its smaller but contemporaneous Burnham-designed neighbor, the Symphony Center, this Beaux Arts beauty should be renamed the

Chicago Temple of Terra-Cotta. Having taught architectural geology for some wonderful students here, I may be biased, but to my mind it's the city's one most splendid structure featuring this clay-based stand-in for stone. The exterior's glazed white cladding, produced locally by the Northwestern Terra Cotta Company, is impressive enough, but a quick detour into the skylighted, two-story atrium reveals just how coolly elegant an atmosphere it can create. It's no second-class simulacrum. Here the white marble of the grand staircase and floor pavers serves to compliment the terra-cotta, and not the other way around.

It must have seemed a foolhardy act to make so white a building in so relentlessly grimy a city. In *The Future of America*, the English novelist and social commentator H. G. Wells noted, at about the time the Railway Exchange was constructed, that

> Chicago burns bituminous coal, it has a reek that outdoes London, and right and left of the [railroad] line rise vast chimneys, huge, blackened grain elevators, flame-crowned furnaces and gauntly ugly and filthy factory buildings, monstrous mounds of refuse, desolate empty lots littered with rusty cans, old iron, and indescribable rubbish. And interspersed with these are groups of dirty, disreputable, insanitary-looking wooden houses—the homes of the people.

This was indeed the time when that bituminous coal, mined from the Pennsylvanian-age strata of the Illinois Basin, was the city's predominant power source, and, as Wells correctly points out, exposed surfaces quickly acquired a dismal coating of carbonaceous soot that was difficult to remove from stone and uncoated brick surfaces. But the Railway Exchange Building's whiteness has the remarkable ability to rise, so to speak, from the ashes. A 1907 photograph in the *Brickbuilder* illustrates why terra-cotta was, in Burnham's era, such a godsend to Chicago architects. The photo, titled "An Enamel Finish Terra Cotta 'Skyscraper' Being Given Its Annual Bath," shows part of this building's Michigan Avenue façade light and sparkling, as intended, and a much darker part still awaiting the arrival of the maintenance crew's suspended scaffold. It certainly demonstrates that one of terra-cotta's chief selling points was its unmatched ability to be cleaned easily and effectively. Building owners no doubt thought a yearly outlay of $1,000—the cited cost for one washing of the entire exterior of the Railway Exchange—was well worth it.

6.8 McCormick Building

332 S. Michigan Avenue
Completed in 1910, northern extension in 1912

Architectural firm: Holabird & Roche
Foundation: Hardpan caissons
Geologic features: Town Mountain Granite, Logan Brick, Roman Brick

The exterior of the lowest two stories here is cloaked in an impressive pink granite. There's also some "black granite" of uncertain identity serving as window trim. The much more extensive and lighter-tinted stone is the "Sunset Red" variety of the Town Mountain Granite, quarried in Marble Falls, Texas, in the midst of a large geologic structure known as the Llano Uplift. This igneous intrusive rock formed from magma that rose from the depths to solidify as a pluton about 1.05 Ga, very near the Mesoproterozoic-Neoproterozoic boundary. This was in the final phase of an immense, worldwide mountain-building event called the Grenville Orogeny. At that distant point in Earth history a supercontinent much more ancient than Pangaea—geologists have dubbed it Rodinia—had come together during a period of unusually vigorous tectonic activity that included various collisions of once-separate landmasses. The Town Mountain Granite has a great deal of interesting detail and visual appeal, as coarse-grained and porphyritic igneous rocks usually do. The larger-than-average crystals in its matrix are pink microcline feldspar. They share the stage with smaller grains of white plagioclase, glassy gray quartz, and black biotite mica.

In contrast to the ruddy granite below, the McCormick Building's upper stories feature an unabashedly plain and unadorned surface composed of buff Logan Brick provided by the Columbus Brick and Terra Cotta Company. That firm's works were located in rural Logan, Ohio, amid the ample clay deposits of the unglaciated southern portion of that state. If you look closely, and preferably with a pair of binoculars, you'll see that the brick here is of the Roman type, which is longer than standard American size. Such Chicago architects as Holabird and Roche, Louis Sullivan, and Frank Lloyd Wright were drawn to the ornamental potential of Roman Brick and used it masterfully in some of their most celebrated projects.

6.9   Chicago Club

85 E. Van Buren Street
Completed in 1929
Architectural firm: Granger & Bollenbacher
Foundation: Rock caissons
Geologic feature: Portland Sandstone

If on a visit to New York City you take a hike through Manhattan's Upper West Side or Brooklyn's Park Slope neighborhood, you'll note that many of the row houses there are made of a striking, maroon-to-somewhat-muddier "brown-stone" that infuses its setting with a sort of homey, autumnal charm. This is the famous Portland Sandstone, a clastic sedimentary rock quarried in the geologically fascinating Connecticut River Valley of southern New England. Were you then to swing by Chicago's Thompson House (site 14.12) or the Near West Side's Church of the Epiphany (site 9.6), you'd be readily forgiven for thinking you're seeing the Portland there, too. But in fact it's the look-alike Lake Superior Brownstone, quarried instead in northernmost Wisconsin and Michigan's Upper Peninsula. In Midwestern cities, certainly including this one, it's a much more common sight than the Portland.

However, in architectural geology the rules of reasonable expectation are often betrayed in diverse and perverse ways. Here at the Chicago Club, certainly the duskiest edifice in this stretch of South Michigan Avenue, it would be eminently logical to think, based on the facts cited above, that its exterior cladding is yet another instance of the Lake Superior stone. But in this case it *is* the Portland. Why did the building's architects choose it instead? Probably simply because the Lake Superior quarry industry was defunct by the end of the Roaring Twenties. After all, the once-popular brownstone look had been deemed unfashionable in Chicago for the previous three decades. As Wisconsin geologist and building-stone expert Ernest Robertson Buckley lamented back in 1898, "Until a few years ago brown stone was all the rage, both for business blocks and residences, but the eye became weary of gazing at long rows of somber colored buildings, and the fashion changed to light colored stone, where it now rests, awaiting the next reversal." It appears the Chicago Club was that next reversal—a very temporary one that in Chicago consisted of this one building and, to my knowledge, no others.

For that reason especially urban geologists should be grateful that this site's designers were willing to buck the current fashion trend. I certainly am. Each time I reach this stop on my Loop tours the pulse quickens, the endorphins come rushing forth, and I risk an acute case of the heebie-jeebies. This is, after all, downtown Chicago's Ground Zero for sedimentary petrology. Each grain of sand encased in this rock offers spiritual redemption and has its own epic history to relate. One section of the stone in particular—it's on the Michigan Avenue side near the corner—bears a magnificent set of ripple marks created when the waves of an early Jurassic lake stirred the sand below them. These ripples were quickly buried by another layer of sand before they could be obliterated. Eventually they turned to stone, where, some 200 Ma later, they were exposed again and set on display in a premium location of a great city 800 miles to the west of where they'd formed. And where they'd formed was quite a place in itself: a great rift valley that signaled the first stage in the breakup of Pangaea and the birth of the North

FIGURE 6.2. Deposited in a rift that formed during the breakup of Pangaea, the Portland Sandstone is famous for its dinosaur footprints and fascinating sedimentary features. One face-mounted slab on the Chicago Club's exterior bears a superb set of ripple marks produced by flowing water 200 million years ago.

Atlantic. As their trackways preserved in Portland beds in the Connecticut Valley testify, early dinosaurs thrived there, where sediments eroded from the igneous rocks in the surrounding highlands were washed into bodies of water ponded on the floor of the rift.

The two main quarrying locations for this stone were in Portland, Connecticut, and upstream in Longmeadow, Massachusetts. For all its popularity in the heyday of its use, the Portland Sandstone soon attracted its share of East Coast detractors because of its habit of peeling and spalling badly. Apparently this was not so much the fault of the stone itself as of how and when it was used. For one thing, it was often installed on building exteriors in face-bedded fashion. In other words, it was cut horizontally in the quarry, parallel to the flat strata there, and then flipped into a vertical position when set onto walls. This provided a greater consistency of appearance, but was contrary to the way the sandstone itself was used to handling stresses. On top of that, some suppliers cut corners by not sufficiently "seasoning" (drying out) waterlogged stone before it was utilized.

The ripple marks on display here have been rotated 90 degrees up from their original flat orientation, and so demonstrate that at least some of the Chicago Club ashlar is indeed face-bedded. But it's more difficult to say whether it was properly seasoned. At any rate, it's easy to find some fairly recently exfoliated patches on the exterior. These unwanted divots are of considerable geologic interest, though. They expose fresh surfaces that reveal the sandstone's true texture and color much better than the older, soot-stained surfaces do.

Formerly, there was one other grand example of the Portland in Chicago: the George Pullman residence in the South Loop's historic Prairie Avenue district. Sadly, that mansion was torn down in 1922, and now the Chicago Club stands alone as the one well-documented Windy City showplace for this fascinating stone.

6.10   Fine Arts Building

410 S. Michigan Avenue
Completed in 1885, 3-story addition in 1898
Architect: Solon S. Beman
Former name: Studebaker Building
Foundation: Shallow and isolated spread footings
Geologic features: Graniteville Granite, Salem Limestone, Scagliola

With a Romanesque affinity evident in its semicircular arches, hefty columns, and rock-faced masonry, the Fine Arts Building is bedecked in its middle stories with limestone provided by the Blue Hole Quarry of Bedford, Indiana—a sure sign that the architectural use of Salem Limestone in Chicago was already a well-established practice by the mid-1880s.

Below it, however, is a much different stone, an igneous rock that is purplish-pink on rough surfaces and a ruddy rose when polished. It's the beautiful if rather repetitively named Graniteville Granite, quarried in the eponymous Missouri hamlet. This locale lies in the St. Francois Mountains, due south of St. Louis and tucked in the heart of the geologic structure known as the Ozark Dome. The rock's parent magma solidified as a pluton under a large volcanic complex a little more than 1.3 Ga, in the Mesoproterozoic era. Here eminently inspectable at ground level, the Graniteville is generally medium-grained, with some of its deep-pink orthoclase crystals considerably larger than the rest—which makes this rock at least somewhat porphyritic. It also contains a great deal of grayish quartz and a lesser amount of biotite and other minerals. The prevalence of orthoclase and the lack of any corresponding plagioclase means that geologists classify this rock specifically as an alkali-feldspar granite.

Over the years, the Graniteville's main selling points have been its vivid coloration and its hard, silica-rich composition. The latter allows it to take and hold a very high polish, and to retain the kind of carved detail visible on this building. And there's yet another attribute: the stone can be extracted in very large and flawless blocks, limited in size only by the lifting power of its quarries' derricks. This meant that it could be fashioned into huge, single-section column shafts, such as those that here adorn the center of the third- and fourth-floor façade. In its late nineteenth-century heyday the Graniteville, often marketed as "Missouri Red," was so popular that it was even a favorite with architects in Vermont and Massachusetts, states that were themselves leading granite producers.

A quick look inside at the main hallway's walls will reward you with an excellent introduction to scagliola, the plaster compound that is painted and polished to resemble choicest stone. Apparently one of its local trade names was "Litho-Marble." As impressive a simulacrum as it really is, it provoked the Chicago correspondent of the Boston-based *American Architect* into one of his more extended assaults on the perceived cultural pretensions of our city. Of this building, he wrote in 1898,

> Its entrance-hall is agreeable enough in general outline, but a disappointment is immediately felt, owing to the fact that it is lined entirely with imitation marble. One of our local publications remarks that, "By the use of litho-marble the architect has obtained the bold and impressive effect of the whole having been carved out of a solid bed of Sienna marble." The bold (and one might also add bad) effect is further heightened by having some lithographs and etchings, such as can be purchased, framed, at almost any department-store for $2.49, hung on these marble walls of the main hall!

Those low-budget lithographs and etchings are now long gone, and any honest eye regarding the hallway will see that the skillful application of scagliola qualifies as its own fine art.

6.11 Auditorium Building

430 S. Michigan Avenue
Completed in 1889
Architectural firm: Adler & Sullivan
Foundation: Shallow, see the description below
Geologic features: Hinsdale Granite, Spruce Head Granite, Salem Limestone, St. Cloud Area Granite, Mexican Onyx Travertine, Scagliola

FIGURE 6.3. Adler and Sullivan's Auditorium Building, completed at a time when the task of safely anchoring ever larger and heavier structures in Chicago's treacherously unstable substrate posed a daunting challenge to the city's engineers and architects. The Michigan Avenue façade, shown here, features two granites, Minnesota's Hinsdale and Maine's Spruce Head, on the lower stories. Indiana's lighter-toned Salem Limestone is used above.

This site, more than any other in this book, merits an explication of its foundation design. For the process of creating a suitable understructure here for everything that rose above it exemplifies how this building's engineers and architects grappled with the formidable challenge of safely constructing a large and heavy structure in Chicago's wet and shifting substrate. It's important to keep in mind that at the time of construction the lakeshore stood just the equivalent of a city block to the east. That meant that the water table—the upper limit of the groundwater—could rise to as little as 12 feet below street level, and as much as 8 feet above the bottom of the foundations. Just how does one perch a building weighing over 110,000 tons on this uninviting bed of waterlogged sediments?

Had the Auditorium been constructed just a half decade later, a network of caissons dug down at least as far as the hardpan would have been the much simpler and ultimately more effective method of stabilization. And ironically it would be Adler and Sullivan, teamed with the preeminent foundation expert

William Sooy Smith, who would introduce this new type of anchoring on a later project, the Chicago Stock Exchange. But without that better technique at hand here, Adler, who was every bit as brilliant an engineer as Sullivan was an architect, devised a complex system of different footing types that has proven its basic soundness ever since. As Adler wrote in the May 1888 issue of the *Inland Architect*, "The design of the foundations of the Auditorium building . . . involved the solution of several perplexing problems, and, as is usual in such cases, this solution implied the adoption of a series of compromises between conflicting positive wants and imperative conditions."

One of those imperative conditions was that the Auditorium's exterior walls were load-bearing. This required a continuous footing of timbers, concrete, and steel rails that was placed under them. In the building's interior, though, the weight from above was borne by cast iron columns set on separate pyramidal minifoundations made of concrete reinforced with rails. But for the area under the great tower on the structure's south flank Adler devised a large raft, at 6,700 square feet in area considerably wider than the tower itself, that is composed of concrete 5 feet deep in which there is embedded a total of eight layers of timbers, steel-rail grillage, and iron I-beams.

The level of meticulousness and foresight that went into this complicated hybrid design can only be marveled at. Smith, who was a consultant here too, had carefully measured the compressibility of the substrate throughout the site by driving piles of varying lengths into it, loading them with containers of water, and observing the settling rates. And Adler, aware that the tower raft would bear more weight than the rest of the foundation system, ballasted it with loads of brick and pig-iron in proportion to how much of the building had been erected at any one time. Then, when the final phase of the construction began and the upper part of the tower rose above the rest of the exterior, the added foundation load was removed at the same rate the tower construction continued upward. While some subsidence was always to be expected with shallow foundations and had been factored in, this painstaking preloading process was intended to prevent the heavier tower section from sinking more than the rest of the structure. Despite all that, the Auditorium Building, which had been set 2 inches above grade to compensate for the originally expected compaction, ultimately did settle much more, and unevenly—up to 28 inches under the tower—for a completely different reason beyond Adler's or Smith's control: after the foundation had been installed with so much care and calculation, the committee overseeing the construction decided to change the exterior from brick to much heavier stone. The settling that resulted from this significant increase in mass can be seen in some of the tipsy, out-of-plumb floors and stairways in the theater section. Fortunately, it did not prove to be a safety risk or threat to structural integrity.

The building's outer surface from the fourth story up is that dependable favorite of the American stone trade, the Salem Limestone, or in architectural parlance, "Indiana Limestone" or "Bedford Stone." Below it and much more accessible at ground level is a harder rock type of similar color, the Hinsdale Granite, on full display here as massive blocks of both rock-faced and dressed-face ashlar. It is also sometimes cited as "Duluth Gray," "Mesabi Heights," or simply "Minnesota Granite." This rock was produced in Hinsdale, Minnesota, a small railroad-siding outpost in the Mesabi Iron-Ore Range. One characteristic of this rock is its large chunks of black hornblende, which quarrymen call "knots." These inclusions are often considered unsightly and undesirable for decorative stone, but apparently they did not offend designer Sullivan's demanding aesthetic sense, because they're common here. Interestingly, though, the company that produced the granite failed to deliver sufficient quantities of the stone on time, and the Auditorium Association ultimately took control of the quarry itself. At one point Adler actually traveled up to the site to smooth out problems triggered by labor unrest—a frequent occurrence in the stone trade in those days.

The Hinsdale Granite is Neoarchean in age, and therefore it's the oldest to be seen on the Auditorium Building. It is composed of large pink orthoclase crystals, gray and white plagioclase and microcline, quartz in small grains, and black hornblende, biotite, and magnetite. But while the stout columns that stand nearby appear to be the polished version of the Hinsdale, they're actually made of the look-alike Spruce Head Granite instead. A product of coastal Maine, it was chosen because, unlike the Hinsdale, it could be extracted in sections large enough to make the single-piece column shafts seen here. Dating to a more recent but still distant episode in geologic history, the Silurian period, the Spruce Head contains sizable knots, too, and is medium- to coarse-grained. Its mineral content includes white oligoclase and microcline feldspars, gray quartz, and black hornblende and biotite.

Yet another granite, probably a later addition or replacement, can be found on the building's rear, facing Wabash Avenue. The bases of the Spruce Head columns are the "Charcoal Black" variety of Minnesota's Paleoproterozoic-era St. Cloud Area Granite. There is also a base course of an unidentified "black granite."

What was originally the hotel section of the Auditorium Building is now home to Roosevelt University. Its lobby interior facing Michigan Avenue contains Sullivan's striking decorative details and still more items of geologic interest. The cladding on the guard station and walls screams for attention: each panel is a showy display of clumps and swirls of pearl white and ochre. The stone producing this visual overload is Mexican Onyx Travertine—a very different travertine than the oft-pitted, off-white, and rather drab Tivoli equivalent that has been much more widely used because of its greater durability and usefulness outdoors. The Mexican Onyx has been quarried in a number of places south of the border, most notably the Tecali area of the state of Puebla, but also in other locales in Puebla, Oaxaca, and Baja California.

The square-sided lobby columns also seem to be clad in some elegant rock type consonant in color with the Mexican Onyx. But their cast iron cores are actually sheathed in a plaster-based marble substitute called scagliola. This technique, in which an artfully pigmented paste is applied to a surface and then polished when hardened, dates back at least to seventeenth-century Italy. It may sound like a cheapskate's stand-in for much more expensive real stone, but it has been used in some of the Midwest's most opulent buildings, and is in fact a product of highly skilled craftsmen (in this case, from the Chicago Art Marble Company) who reproduce the complicated patterns and luster of real stone with uncanny precision.

6.12   Equestrian Indian Statues

Grant Park (S. Michigan Avenue and E. Congress Parkway)
Completed in 1928
Sculptor: Ivan Meštrović
Geologic features: Cape Ann Granite, Bronze

Also known as the Spearman and the Bowman, these statues stand as the mounted guardians of this grand southern entrance to the Loop. The tensed monumental figures are made of bronze that, unlike the lions of the Art Institute, have not been allowed to weather into the typical bluish-green patina. Instead, they have a handsome deep-brown coating more characteristic of one form of copper oxide. This effect, often used for bronze statuary, is usually created artificially by the application of patina-producing chemicals.

The pedestals below are also worth a close look. They're clad in the rarer, faintly greenish-tinted version of Cape Ann Granite. Marketed as "Rockport Green," this Silurian stone was quarried in Massachusetts along the Atlantic seacoast northeast of Boston. The polished surfaces show some staining, but in many places still beautifully reveal the medium- to coarse-grained matrix of light-gray orthoclase feldspar, olive quartz, and black biotite and hornblende.

6.13   Clarence F. Buckingham Fountain

Grant Park, just east of the Congress Parkway terminus
Completed in 1927
Architect: Edward H. Bennett
Artist: Marcel Francois Loyau
Engineer: Jacques H. Lambert
Geologic features: Murphy Marble, Graniteville Granite, Bronze

Designed by the architect also responsible for the Michigan Avenue Bridge Houses, this fountain is the centerpiece of Grant Park's symmetric Beaux Arts formality. With the incomparable Chicago skyline rising behind it, it couldn't have a more magnificent backdrop. Each hour during its operating season it circulates over 14,000 gallons of water that shoot in graceful arcs from 134 jets, including those issuing from the mouths of sculptor Loyau's cavorting horsey-fish hybrids. These fantastic creatures are bronze and have weathered, as that alloy is often allowed to do for ornamental and protective reasons, to the classic bright-green patina of copper carbonate, copper sulfate, or both.

The fountain's 280-foot-wide bottom pool and the three basins that stand above it are mostly composed of the "Etowah" variety of the Murphy Marble, quarried in the vicinity of Tate, Georgia. It first formed as limestone in the Cambrian or Ordovician, on the continental shelf at the edge of the now long-vanished Iapetus Ocean. Later it was metamorphosed into true marble during the three episodes of mountain building that occurred in eastern North America later in the Paleozoic era. Its distinctive pink color and its patterning of darker folds and foliations have made it a favorite of Chicago architects since at least the 1880s. While the black to silver-gray markings are quite certainly due to the stone's graphite content, the source of the overall pinkness has been a matter of debate, with some geologists surmising it is the iron-bearing mineral hematite that caused it, and at least one other stating that trace amounts of manganese are probably responsible instead. In any case, the bulk of the marble here is calcite.

Depending on when you visit the fountain, you might also notice that some of the Murphy sports an unsettling lime-green color instead. This is due to a temporary coating of calcium-loving algae that thrive in this wet, sunny location. Seemingly immune to this problem are the columns that support the uppermost basin. They have been fashioned from the Graniteville Granite taken from the A. J. Sheahan quarry in southeastern Missouri. This ruddy rock selection, actually an alkali-feldspar granite, predates the Paleozoic origins of the Murphy Marble. In fact, it's more than twice as old, having solidified from magma in the Meso-proterozoic era.

6.14 DePaul University Lewis Center

25 E. Jackson Boulevard
Completed in 1917
Former name: Kimball Building
Architectural firm: Graham, Anderson, Probst & White
Foundation: Hardpan caissons
Geologic features: Conway Granite, Terra-Cotta, Ceramic Tile

The main element of this building's exterior is handsome white terra-cotta, but at the façade's base there is also some notable stone cladding. As one approaches, it becomes easier to see that it's a granite that offers the pale suggestion of greenishness. That's a sign that it is either a rarer variety of Silurian-period Cape Ann Granite from Massachusetts or the "Redstone Green" version of New Hampshire's Conway Granite, which dates to the Jurassic. In this case it's the latter, but the two rock types can be difficult to distinguish because both tend to be coarsegrained and owe their unusual tint to their yellowish-green quartz crystals. The other main constituents of the Conway Granite are biotite and hornblende, each essentially black, and gray orthoclase.

The ground-floor interior is also worth a quick look. The wall cladding is tan, glazed faience—ceramic tile of uncertain origin that makes the perfect complement to the terra-cotta, the other clay-derived material.

6.15   The Alfred

30 E. Adams Street
Completed in 1925
Former name: Hartman Building
Architect: Alfred S. Alschuler
Foundation: Wooden piles
Geologic feature: Marcy Metanorthosite

Yet another faintly green-tinted crystalline rock awaits you by the main Adams Street entrance of this building. The preceding description of the Lewis Center might lead you to conclude that it's either the Conway Granite or the Cape Ann Granite. But the stone here is actually a third and much rarer type, and one that isn't even granitoid. And so begins another adventure on the misty moors of building-stone identification.

The Coldspring 1960s inventory of Chicagoland's decorative stone describes the cladding here as "Moss Green (New York)," which is odd because that trade name is usually applied to a true granite quarried in Kenora, Ontario, though now it's better known as "Kenoran Sage." But the Coldspring compiler took pains to identify this stone's locale of origin as being south of the Canadian border, and specifically in the Empire State. That and some mineralogical clues suggest that it's actually the stone marketed nowadays as "*Mountain* Green," or what geologists call the Marcy Metanorthosite. When polished, the stone types from both countries look quite similar, but the Marcy contains minute specks of deep-red garnet, and tends to show weak foliation. This means that it often bears evidence of having its minerals arranged, at least here and there, in a loosely linear or banded orientation. On very close examination, the cladding here, though at time of writing not in the best shape, does seem to have some garnet. And there is, unless I'm guilty of wishful thinking, a subtle striped pattern to the rock. It's hard to pick out at first, until you tilt your head 90 degrees and see the bands of light and dark minerals running from top to bottom.

The *meta* prefix tagged onto the igneous rock term anorthosite indicates that the stone has experienced enough alteration by heat or pressure in its long history to now qualify as a metamorphic rock instead. The Marcy is quarried in the town of Au Sable Forks, in New York's Adirondack Mountains—a scientifically enigmatic region because geologists aren't sure why it's still being uplifted, albeit at the modest rate of 3.7 millimeters (0.15 inch) a year. This rock came into being at approximately 1.1 Ga, during the first of eastern North America's great mountain-building events, the Grenville Orogeny, and the assembly of the supercontinent Rodinia. Anorthosite, whether metamorphosed or not, is a rather uncommon rock at the Earth's surface, but in the Adirondacks it's abundant. It completely lacks quartz, which is a necessary ingredient of granite, and is mostly just plagioclase feldspar (here the greenish-brown grains) and pyroxene (the black ones). The additional garnet content is a special characteristic of this particular stone type.

6.16   Palmer House Hotel

17 E. Monroe Street
Completed in 1927
Architectural firm: Holabird & Roche
Foundation: Rock caissons
Geologic features: Quincy Granite, South Kawishiwi Troctolite, Salem
    Limestone, Bronze

The focal point of this site description is on the Wabash Avenue side of this famously opulent hotel. The canopy, doorways, trim, and neoclassical ornament of the adjacent storefront façades are made of bronze treated to keep its original metallic sheen and color intact. No brown or green weathering products would be appropriate here.

This massive building's lowest six stories are clad in Salem Limestone, and along the Wabash ground-story façade there are visible fragments of Mississippian marine-invertebrate fossils. But two other types of stone, both igneous, are also present and they stand out in direct contrast to the Salem's cheerful buff color. The first, framing the hotel entrance, is the Quincy Granite, formerly quarried in the Massachusetts city where the locals very quickly point out to strangers that the correct pronunciation is Kwin-Zee, not Kwin-See. Most "black granites" aren't granite at all, but this deep bluish-gray rock is—albeit the special type known as an alkali-feldspar granite. What sets it apart from normal granites is the fact that it's completely devoid of any type of plagioclase feldspar. Instead, it's composed of 60 percent microcline (an alkali feldspar), about 30 percent quartz, with the rest being hornblende and trace minerals. Long thought to date to the late Ordovician or early Silurian, the Quincy has recently been more definitively assigned to the early Devonian, based on recent highly accurate U/Pb (uranium and lead isotope) zircon dating.

The other dark stone is used here for the low base course that meets the sidewalk. This is the South Kawishiwi Troctolite, quarried in Minnesota's Superior National Forest. Its large, labradorescent plagioclase crystals gleam amid the stains and scuff marks. By far the oldest rock of the three discussed in this section, it dates to the Mesoproterozoic.

6.17 Century Building

202 S. State Street
Currently known by its street address
Former name: Buck & Rayner Building
Completed in 1915
Architectural firm: Holabird & Roche
Foundation: Rock caissons
Geologic features: Larvikite Monzonite, Terra-Cotta

The terra-cotta cladding on the bulk of the exterior here is best seen from across the street. What can be observed in much greater detail is the main entrance's Larvikite Monzonite trim, which simply must be a later addition. It's the "Blue Pearl" variety of this showy stone, the undisputed king of labradorescence. Permian in age, it's quarried near Larvik, Norway, and owes its glittery nature to its perthite content. Black pyroxene is also present.

6.18 Lytton Building

14 E. Jackson Boulevard
Current name: Richard M. and Maggie C. Daley Building of DePaul
    University
Completed in 1913
Architectural firm: Marshall & Fox
Foundation: Caissons (type not specified)
Geologic features: Cape Ann Granite, Sylacauga Marble

Now part of DePaul University's downtown campus, this building was originally the Lytton & Sons department store. The exterior's basal stories are clad in the "Rockport Gray" variety of the Cape Ann Granite, a light-gray stone particularly in favor with Chicago architects in the 1910s. Dating to the Silurian period, the Cape Ann comes from the North Shore—the North Shore of Massachusetts, that is. It's composed primarily of white orthoclase, gray quartz, and black hornblende, and its lack of plagioclase makes it specifically an alkali-feldspar granite.

An even greater treat can be found in the building's interior, and it's especially easy to spot if you use the Jackson Boulevard entrance. (It's no doubt wise to let the DePaul security guard on duty here know why you're examining the decorative stone so intently, especially if you lack the self-assured and worldly aura of a currently enrolled DePaul student.) At the end of the long and narrow lobby rises a handsome staircase made of gleaming, white and gray-veined Sylacauga Marble. Also known as "Alabama Marble" in recognition of its state of origin, it was quarried in the Talladega County town of Sylacauga. The US Geological Survey lists its parent formation's age as "Cambrian?-Ordovician?" That may sound frustratingly tentative, but then some rock units lack the necessary evidence and just can't be pinned down to a single period unequivocally. But this can be said without hesitation: this lustrous and lovely marble, with a recorded calcite content of 99.4 percent, is as pure and purely beautiful as the choicest Italian Carrara equivalent. Described as somewhat finer in grain than its equivalent from Vermont, and considerably more so than Georgia's coarse-textured Murphy Marble, it's every bit as select and desirable for indoor architectural use as any fancy foreign import sporting a more extensively cultivated reputation.

Beneath the Sylacauga is a band of nicely contrasting trim. This darkest-green and white-veined stone was apparently not documented, but it's clearly a serpentinite or ophicalcite.

6.19   Harold Washington Library Center

400 S. State Street
Completed in 1991
Architectural firm: Hammond, Beeby & Babka
Geologic features: Southern Swedish Red Granite, Carrara Marble,
    Aluminum

One would think that the more recently constructed a building is, the more surviving documentation on its building materials there'd be. But here and elsewhere in the city that simply isn't so. It seems the golden age of recording and retaining specifics ended somewhere in the first half of the twentieth century.

A multitude of sources cite the Washington Library's red-brick and red-granite exterior, but none mention its materials' origins. After many hours spent checking with the archivists of the Chicago History Museum, the Art Institute, and the library itself, I was no closer to my goal of pinning specific names on stone and brick. Fortunately, though, I did get a reply from the building's designers—a rare enough result when contacting architectural firms large or small, and therefore especially appreciated in this case. The person who assisted me reported that the plans for the building had been donated to one of the institutions cited above; but subsequently they could not be found there. However, she and her colleagues did discover one surviving design sketch in their office. It indicated that the architects' first choices for the exterior granite had been selections known as "Napoleon Red" and "Imperial Red," while "Carnelian" would do if the first two weren't available. My study of the exterior stone strongly suggests one or both of the top picks were in fact used. Each is a variety of what I call Southern Swedish Red Granite, a rock type in special favor with postmodernists since the 1980s. The "Carnelian," which is a brand of South Dakota's Milbank Area Granite, tends to be smaller-grained than the stone here. If my assessment is correct, these rock- and ashlar-faced blocks date to the Proterozoic eon. Southern Swedish Red varieties are quarried in several different localities in that Norse country's bottom quarter, and their individual radiometric dates vary by as much as several hundred million years.

While the brick's source remains unknown, you might think the oversized acroteria are more easily identified because of their light-green color, which is a sure sign, usually, of bronze sporting its full copper-salts patina. But they're not that noble alloy; rather, they're aluminum painted to simulate it. This makes one wonder, given that metal's featherweight properties, if these mannerist ornaments are likely to blow away when the next derecho-force winds hit downtown—as some Chicagoans no doubt wish they would. (Personally, though, I like their nose-thumbing hypertrophy a little bit more each time I see them.)

A quick trip inside the library via the State Street entrance reveals more interesting use of decorative stone. The hallway wainscoting is a real screamer, and what I take to be, once again without surviving provenance, a wildly brecciated form of Carrara Marble. If it isn't that, it's an astoundingly good imitator of it. To many, the mention of Carrara stone conjures up images of flawlessly white, Statuario-grade stuff, but it actually comes in a remarkably large assortment of patterns and shades. This one most closely resembles the variety known for centuries as "Pavonazzo." But regardless of the trade name it goes by, the Carrara began its existence as Jurassic limestone precipitated in the Tethys Ocean, the great precursor to today's Mediterranean Sea. Only much later, in the Oligocene, did the forces of mountain building transform it into what would become the world's most famous and sought-after marble.

6.20   Chicago Bar Association

321 S. Plymouth Court
Completed in 1990
Architectural firm: Tigerman, McCurry
Geologic features: Buddusò Granite, Concrete, Aluminum

The Chicago Bar Association's headquarters may be tucked into one of the Loop's more easily overlooked crannies, but it's brimming over with geologic significance. The building's frame and most of its exterior is composed of what one could reasonably call limestone, were it qualified with the highfalutin scientific tag *anthropogenic*—that is, manmade. This synthetic rock is concrete, that wonder substance with a remarkably long record of human use. It's made by mixing cement (nowadays usually the Portland variety) with aggregate and water. In the fascinating chemical reaction that results, the cement's lime (calcium oxide and calcium hydroxide derived from the burning of calcium carbonate in natural limestone) is converted back into calcium carbonate, but this time in a form that's most easily moved, manipulated, and molded to the desired human use.

On the building's lowest two floors the concrete is sheathed with a light-gray Sardinian rock selection, the Buddusò Granite. This stone type is one of a small group of European granites that became all the rage with Chicago's postmodernist architects, who've unwittingly increased the geologic diversity of the city quite significantly by eschewing very similar offerings from American quarries. And who can deny that the Buddusò, a major export of the large and rugged Mediterranean island lying to the west of mainland Italy, is a very apt and attractive choice here? This coarse-grained igneous intrusive rock, technically termed a monzogranite, formed from magma that cooled beneath the Earth's surface in the late Carboniferous or Permian period, during the Variscan mountain-building event associated with the assembly of the supercontinent Pangaea.

In this neighborhood goddesses abound. Three blocks away, atop the Chicago Board of Trade, proud Ceres stands above all; here, hovering above the main entrance, is Mary Block's sculpture of Themis, the ancient Greek deity who in one form or another often hangs out in courthouses and other lawyerly locales. No doubt this is because she's the personification of justice and wise counsel. Like

Ceres and the spires that rise from this structure's roof far above, she is made of aluminum, the most abundant metal and third most common element in the Earth's crust. Almost all aluminum in use today comes from the ore known as bauxite, a rock that forms from the orange- or red-tinted clay deposits known as laterites that are characteristic of humid, tropical climates.

THE LOOP

Southwestern Quadrant

7.1 Harris Trust & Savings Bank

111 W. Monroe Street
Completed in 1911
Architectural firm: Shepley, Rutan & Coolidge
Foundation: Rock caissons
Geologic feature: Stony Creek Gneiss

With four immense Ionic columns that float majestically above the ground story, the main façade of this bank building reiterates a theme frequently seen in the LaSalle Street area—that of solidity and stability. This is meant to be a holy temple of commerce and capital. However, the reddish Stony Creek Gneiss cladding here is anything but staid and sedate. This becomes especially apparent when you examine the building's blank western wall, which is surfaced in large panels of this striking architectural stone. Note the striping and curves and swirls that denote foliation, or the linear orientation of minerals, absent in unmetamorphosed granites. This rock has clearly been through a lot in its long history; its patterns suggest tectonically triggered movement and flow. Make sure you take the time to examine its suite of minerals, which includes pink to brick-red microcline feldspar, gray quartz, and white oligoclase feldspar. Bits of black biotite and silvery muscovite, both micas, are also visible.

Quarried in the vicinity of Branford, Connecticut, near the shore of Long Island Sound, the Stony Creek is renowned for its use in the base of the Statue

FIGURE 7.1.   To the geologist this seemingly blank wall at the Harris Trust is a well-stocked storybook. The cladding is composed of polished slabs of Connecticut's Stony Creek Gneiss. Its dark, undulating banding and foliated texture reveals that this rock, often described as a granite, has in fact undergone considerable metamorphosis.

of Liberty and Manhattan's Grand Central Station. Chicago boasts some fine examples of it too, including the pedestal of Jackson Park's Statue of *The Republic* and River North's Newberry Library. But the Harris Trust is the place to see why, even though it's always marketed as a granite, it's formally classified as a granite gneiss—that is, a gneiss that certainly once was granite but is now altered enough to be considered unequivocally metamorphic.

In some spots in its outcrop area the Stony Creek also takes on the characteristics of a migmatite, a "mixed rock" that contains both the older, probably Neoproterozoic gneiss seen here at the Harris Trust and younger unaltered granite, probably Permian in age, as well. A migmatite is a vivid demonstration of nature's impish tendency to subvert human attempts at neat classification and clear distinctions. Is it igneous? Metamorphic? Actually, it's both, a rock that in this case has experienced at least three phases of character-altering development. In the Stony Creek's case, the story began when a body of magma was emplaced in the crust of the wandering microcontinent of Avalonia, then situated in the Southern Hemisphere. The rock was subsequently changed into gneiss, perhaps as a result of the great tectonic mayhem produced when this peripatetic terrane collided with ancestral North

America. And then, much later yet, fresh granite-forming magma was injected into it, here and there, as Africa collided with New England to form Pangaea.

7.2   Marquette Building

140 S. Dearborn Street

Completed in 1895

Architectural firm: Holabird & Roche

Foundation: Shallow and isolated spread grillage footings; hardpan
   caissons added under the eastern outer wall in 1940

Geologic features: St. Cloud Area Granite, Carrara Marble, Porter
   Brick, Tiffany Enameled Brick, Northwestern Terra Cotta, Bronze,
   Favrile Glass

The Marquette Building is unquestionably one of the crown jewels of Chicago architecture, and it's another one of the Loop's top-tier displays of geologically derived materials. My only problem with this site I love so much is that it tends to gobble up my tour groups, who twenty minutes after they enter the lobby are still there, cooing and purring and gazing in epiphanies of admiration. It's very difficult to get them disgorged back onto South Dearborn in a timely fashion.

But what a lot there is to see, starting with the distinctive chocolate-brown exterior. At some distance it could be taken for brownstone cladding, but in fact the effect here is achieved instead by a combination of darkly tinted terra-cotta and brick. The former material was fashioned in Chicago by the Northwestern works; the latter was produced in Porter, Indiana, by the Chicago Hydraulic-Press Brick Company. For the building's inner light well, not usually on view to the public, reflective white enameled brick was used. It was made by the Tiffany Company of Momence, Illinois. Unlike glazed brick, which has a decorative coating that does not fully penetrate into the burnt clay beneath it, enameled brick's coating is fired at a higher temperature so that it fuses more thoroughly with the clay. This makes it considerably more resistant to crazing and spalling.

In contrast, the main entrance on Dearborn features an effective combination of three stone-clad, two-story pilasters and four hefty bronze panels that depict scenes from the life of the man honored in the building's name: French Jesuit explorer Jacques Marquette. Bronze was also used in the magnificently ornamented doorways. And the polished exterior stone is one of the most popular selections of the very widely used St. Cloud Area Granite family. It's been marketed as "Rockville" and "Rockville Pink." Quarried in central Minnesota and Paleoproterozoic in age, it's very coarse-grained with a striking blend of pink feldspars (mostly orthoclase but also a fair amount of microcline), gray quartz, and black biotite.

Once in the door, it's easy to see why my tour companions get so easily trans-fixed. Hovering above the inner lobby is a wraparound mosaic of Tiffany Favrile Glass with more Marquette scenes. (This makes the Marquette Chicago's Build-ing of the Two Tiffanies). The vibrancy of the mosaic's colors always reminds me of an epiphany of my own from many years ago, when I beheld the great apse mosaics of the sixth-century Basilica of Sant'Apollinaire in Classe, Italy. Here at the Marquette, as at the Cultural Center and Marshall Field's, it's good to see this ancient Roman and Byzantine mode of art so well represented a millennium and a half later in an American city.

Less immediately eye-catching in the same space is the polished stone that provides a tempering elegance. Straight from the Apuan Alps, it's the paragon of all marbles, the Carrara. Originally a Jurassic limestone in the bed of the great Tethys Ocean, it was later metamorphosed into its present beautiful form in the Oligocene. While the Carrara has always faced stiff competition from almost identical American marbles quarried in Vermont, the Old South, and Colorado, its unmatchable reputation that stretches back two thousand years has always made it a favorite choice of Windy City architects.

7.3   Chicago Federal Center Plaza

140 S. Dearborn Street
Completed in 1974
Architects: Ludwig Mies van der Rohe; Schmidt, Garden & Erikson;
    C. F. Murphy Associates; A. Epstein & Sons, associate architects
Geologic feature: St. Cloud Area Granite

Each day this Spartan but spacious plaza brings welcome relief from the shaded confinement of the urban canyons to thousands of office workers and tourists. It's also the perfect supplement to the previous stop at the Marquette Building just across Adams Street. There the entrance is graced with polished cladding of the "Rockville" variant of the Paleoproterozoic-era St. Cloud Area Granite; here the large square pavers are the same rock, more muted in its colors because it's in unpolished form. Its finish may in fact be flamed or thermal; in other words, it may have been treated with a blowtorch to better seal it and provide a less slip-pery surface. Alternatively, it may have been mechanically honed or subjected to some other finishing process suitable for pavers intended to be as enduring, watertight, and skid-resistant as possible.

7.4   Palmer Building

27 W. Adams Street
Completed in 1872

Architect: C. M. Palmer
Geologic features: Milbank Area Granite, Cast Iron

Now the western portion of the famous Berghoff Restaurant, the Palmer Building is one of the very few surviving places in the Loop that have an ornamental
cast iron façade (in contrast to the more widespread use of the same material as
an internal load-bearing element). The cast iron here is an early example of what
today might grandiosely be called "modular prefab facing units." This is a classic
nineteenth-century architectural application of the metal that is second in abundance in the Earth's crust only to aluminum. It's strong when compressed, weak
under tension (i.e., when pulled apart); it starts to lose its structural integrity at
about 800 degrees Fahrenheit, which is well below its melting point of over 2,000
degrees. This last trait was amply demonstrated during the Great Fire of 1871,
when earlier buildings' cast iron façades failed in the flames along with highly
flammable timber frames and heat-shattered stone.

At ground level there is, to the delight of the rockhound, what must be a much
more recent addition: a damp course and trim of the "Dakota Mahogany" variety
of eastern South Dakota's Milbank Area Granite. This gorgeous reddish-brown
Neoarchean stone, here in polished form, has an age of at least 2.5 Ga. Consequently it's one of the oldest rock types to be seen on this city's streets. It owes its
coarse texture and rich coloration to its large grains of pink orthoclase feldspar,
gray quartz, and black biotite mica.

7.5   Stone Building

17 W. Adams Street
Completed in 1872
Architect: Unknown
Geologic features: Berea Sandstone, Milbank Area Granite

The aptly named Stone Building constitutes the eastern portion of the Berghoff
Restaurant, and unlike its neighbor, the Palmer Building, it features an upper
façade not of cast iron but of a buff-colored rock type architectural historians
call, a little vaguely, "Cleveland Sandstone." In fact it's almost certainly the Berea
(buh-REE-uh) Sandstone offered by the Cleveland Quarries Company and its
various predecessors. Once thought to be Mississippian in age, it has now been
docketed just a little bit below that on the stratigraphic column, at the very top of
the Devonian. This fine-grained clastic sedimentary rock, outcropping in Ohio's
northernmost tier of counties, was originally sought out for use as millstones.
By the time of this building's construction, though, it had become widely popular
in architectural settings. Nowadays it can often be found gracing the exteriors of
some of the grandest of old Midwestern courthouses, including one in the Illinois

River town of Ottawa, a ninety minutes' drive from the Berghoff's front door. Under a microscope one discovers that this rock's sand grains are subrounded to quite angular, rather than more fully spherical, as in many other commercially valuable sandstones. This accounts for both the Berea's characteristic abrasive texture and its time-honored alternative name, the "Berea Grit."

The damp course of Milbank Area Granite found on the Palmer Building section of the restaurant continues straight across the ground story here.

7.6 Monadnock Building

53 W. Jackson Boulevard

Completed in 1891 (original, northern building: Monadnock and
    Kearsarge sections), 1893 (southern addition: Wachusett and
    Katahdin sections)

Architectural firms: Burnham & Root (1891), Holabird & Roche (1893)

Foundation: Shallow and isolated spread footings of the grillage type;
    caissons under west wall added in 1940

Geologic features: Graniteville Granite, Chicago Anderson Pressed
    Brick, Molded Brick, Tiffany Pressed Brick

The Monadnock Building is known throughout the architectural world for the stark power of its design and for its load-bearing masonry exterior, despite the fact even its two original portions derive some of their support from interior cast iron columns and beams of wrought iron. Still, the ground-level walls, 6 feet thick at their base on the original sections, are eloquent testament to the crucial structural role and the primal beauty of its brickwork. This stolid edifice is a fitting human analog to the great New Hampshire mountain from which it takes its name. Both Monadnocks are geologic features of the first order.

Designed shortly before the advent of the deep caisson foundation, the Monadnock Building has a system of shallow footings similar to that used in another architectural heavyweight of the era, the Auditorium Building. And like the Auditorium, it's a little less visible than it used to be. Originally it was set 8 inches above grade to compensate for settling; in fact there's been about 20 inches of subsidence despite the care taken to deal with the oozy substrate.

Framing the building's entrances is a stone that provides a brighter contrast to all the surrounding somberness. This was, like the brick itself, a favorite choice of designer John Wellborn Root—"Missouri Red," known to geologists as the Graniteville Granite. This Mesoproterozoic intrusive rock, taken from the heart of the Ozarks' St. Francois Mountains, can also be found on Root's Rookery. But here its surface has not been polished. Rather, it was peened to produce a roughened, pitted surface that gives it a lighter tint than usual. Its mineral composition features a familiar cast of characters: pink orthoclase, gray quartz, black biotite.

——————— C H I C A G O ———————
Anderson PRESSED BRICK *Company*
MANUFACTURERS OF
—PLAIN AND ORNAMENTAL PRESSED BRICK.—

FINEST QUALITY OF PRESSED BRICK
FOR BUILDING FRONTS.

OFFICE, 157 LA SALLE STREET, CHICAGO, ILL.
SEND FOR ILLUSTRATED CATALOGUE.

FIGURE 7.2. This 1886 *Inland Architect* advertisement provides a bird's-eye view of the Anderson works sited on the west bank of the Chicago River, at what is now the intersection of W. Webster and N. Elston Avenues. Here the brick for the Rookery and the earliest section of the Monadnock Building was produced.

The maroon brick of the earliest, northern portion of the building has a certain family resemblance to that of the Marquette Building just up the street on Dearborn, but it was manufactured by a major local firm, Chicago Anderson Pressed Brick. Most likely, it's Anderson's "Brown Obsidian" variety, which Burnham and Root had already put to such good use in the Rookery. This company employed the hydraulic production technology pioneered in St. Louis, and this state-of-the-art technique provided dense, rock-solid, and also custom-molded brick that measured up to both this site's unforgiving structural demands and the graceful curves of Root's design. But a resounding note of praise is also due to the world-class masons supplied by the general contractor, George Fuller & Company. They made the myriad details of the architect's vision all come together so superbly.

The brick of identical hue on the Monadnock's southern, internal-framed addition could easily be taken to be an Anderson product, too, but it was made instead by an up-and-coming competitor, Tiffany Pressed Brick of Momence, Illinois. A 1907 company history notes that its brown bricks were made by mixing manganese into the local clay normally used for its standard red line. At first the Tiffany operation relied on clay dug on its own property east of Kankakee, but as its range of colors expanded it also imported clays from nearby Grant City, which when fired turned pink, and from the suitably named Clay City, Indiana, which produced a buff color.

7.7　Fisher Building

343 S. Dearborn Street
Completed in 1896
Architectural firm: D. H. Burnham & Company
Foundation: See the description below
Geologic feature: Northwestern Terra Cotta

Happily, terra-cotta cladding and ornament is a common sight in Chicago, but nowhere has it received a more playful treatment than here, where its unglazed version is on display. The fired-clay sections, which were crafted on Chicago's North Side by the Northwestern Terra Cotta Company, are an unusual scheme of orange and tan that in places is also splotched dark brown, like snakeskin ready to writhe and wriggle. The façade is populated with vegetative motifs, an aquatic bestiary, bambini, and other fabulous figures set in a Gothic framework. Besides providing a fireproof, lightweight, and relatively inexpensive alternative to stone, terra-cotta is protean. It can assume practically any tint, pattern, or form floating in its designer's imagination. But like all other materials exposed to heat, wind, cold, and water, it and the metal fasteners securing it are doomed to deteriorate, sometimes dangerously, with the passage of time. Major restoration and replacement of the original units was done here in the first decade of this century.

The Fisher's foundation is a curious one, and embodies a doubly cautious strategy of both shallow and deeper anchoring. Beneath the building's isolated spread footings, in earlier construction projects deemed sufficient in themselves, project 25-foot wooden piles that add extra stability. This unusual approach was tried at a time when the superiority of the newly devised caisson system had not yet become fully apparent. But it shows that, with the ever-increasing height and weight of skyscrapers at the end of the nineteenth century, the always suspect reliability of shallow foundations had reached its absolute limit. In fact, despite all the precautions taken the Fisher Building did ultimately lean somewhat.

7.8　Old Colony Building

407 S. Dearborn Street
Completed in 1894
Architectural firm: Holabird & Roche
Foundation: See the description below
Geologic features: Salem Limestone, Winslow Junction Brick, Roman
　　Brick, Terra-Cotta

With striking tower bays projecting out from its four corners, the Old Colony holds its own as a revered architectural landmark in a neighborhood that also boasts the incomparable Monadnock and Fisher Buildings. Its foundation is a series of shallow spread footings of the grillage type. However, an alarming amount of subsidence under its shared southern wall prompted the quick addition of four caissons there under the personal direction of that great guru of foundation engineering, William Sooy Smith. But as was often the case with tall buildings resting mostly or entirely on shallow footings, the Old Colony later experienced significant sinking over all of its footprint. In fact, in the first nine months of its existence there was an average settling of just over 4 inches.

Above ground, the building is a handsome succession of the "Blue Bedford" variety of Salem Limestone, nicely carved in places, on the lowest three floors; and above it a buff-colored pressed brick that at first glance might seem to be the well-known Cream City type produced in Milwaukee. That's not a bad guess—the Cream City is a fairly common sight in greater Chicagoland—but in fact it's Winslow Junction Brick, made in that New Jersey locality for the Philadelphia-based Eastern Hydraulic Press Brick Company. At Winslow Junction the brick's raw material was taken from the Cohansey Formation, an extensive deposit of unlithified sand, gravel, and clay that has been tentatively assigned to the Miocene epoch. As a 1904 State Geologist's report noted, "the Cohansey clay is often found to make a good buff-burning brick."

Careful examination of the Old Colony's upper-story Winslow Junction Brick reveals that its dimensions are rather nonstandard: at about 12 inches, it's noticeably longer than normal. In other words, it can also be classified as Roman Brick, certainly not because it was produced in the Eternal City but because its shape is reminiscent of one of the standard brick shapes widely used across their empire by ancient Roman architects. Here it shares the building's upper façade with complementing segments of white terra-cotta trim.

7.9   Chicago Board of Trade Building

141 W. Jackson Boulevard
Completed in 1930
Architectural firm: Holabird & Root
Foundation: Rock caissons
Geologic features: Salem Limestone, St. Cloud Area Granite, Belgian
    Black Limestone, Uchentein Breccia, Boone Limestone, Aluminum

The Board of Trade Building, on its dramatic perch at the foot of LaSalle Street, obeys what I call the Grand Art Deco Formula: a skyscraper as mountain of Salem Limestone, in one sense massive and blocky yet also graceful in its setbacks, garnished at the bottom by a crystalline stone that offers rich color and detail in contrast to the great sweep of khaki blandness above it. And at its summit stands a supreme use of our small planet's most common metal, aluminum, in the form of John H. Storrs's 31-foot-tall abstraction of Ceres, the Roman goddess of agriculture, who's all the more divine for being faceless. To my eye, she's the best sited statue in the city, and at her great height the most truly Olympian in aspect.

Every time I stand by the sidewalk bollards just across Jackson from this building and gaze goddessward at all that stone, I'm struck speechless by the dazzling if ultimately temporary defiance of gravity that one species of the Great Apes has achieved here. Consider how much of an ancient sea bottom, a world of carbonate shoals, coastal lagoons, and tidal channels that constituted southern Indiana in the Mississippian, has been thrust up into the air here; how many myriads of marine creatures rest suspended in their calcite tombs high above the streets of a modern city. The sheer quantity of former life this cliff face holds is harrowing. We have become many things in our slim history, among them, a tectonic force.

At the Board of Trade's sister-in-design, 333 N. Michigan, the crystalline rock of the building base is the anarchic, swirling Morton Gneiss. Here it's a more even-tempered rock type, a variety of St. Cloud Area Granite marketed as "Cold Spring Pearl Pink." It was quarried from one of the bodies of felsic magma that rose from the depths and hardened in the Paleoproterozoic era to create the Minnesota Batholith. That happened five times farther back into eternity than when the Salem formed in its tropical shallows. In Chicago it's not the most common brand of St. Cloud to be seen, but it may very well be the most beautiful. From a few paces away its matrix of coarse crystals—black hornblende and biotite, glassy gray quartz, pink and white feldspars—resemble the colored dots of a pointillist painting. An essay in dynamic equilibrium, they create the pleasant illusion of maintaining overall stasis while vibrating slightly in Brownian motion.

As is usually the case with Chicago's Art Deco skyscrapers, the ground-floor lobby here offers a variety of fascinating and visually impressive exotic rock types. Purportedly eleven different "marbles" were used. Those documented are the veined Belgian Black Limestone, the "Escalette Breche" variety of the pinkish-tan Uchentein Breccia from the French Pyrenees, and floor pavers of buff, Mississippian-age Boone Limestone ("Batesville Marble"). The last of these, quarried in northeastern Arkansas, is quite similar to the much more common Salem Limestone of Indiana, but it has the added virtue of taking a high-gloss polish. Certainly the real show-stopper here is the Belgian Black that adorns the massive, stylized piers. It is either Devonian or Carboniferous, depending on exactly which strata it was taken from.

7.10   Federal Reserve Bank

230 S. LaSalle Street
Completed in 1922
Architectural firm: Graham, Anderson, Probst & White
Foundation: Rock caissons
Geologic feature: Salem Limestone

The architectural order on display in this stolid exemplar of neoclassicism is the Corinthian, with column capitals replete with acanthus leaves and small volutes. In common with the Illinois Merchants Bank and most of the Board of Trade, the exterior here is Salem Limestone. A feature in a 1922 issue of the trade journal *Stone* notes that it was specifically extracted from the Dark Hollow Quarry, just northwest of Bedford, Indiana.

7.11   Illinois Merchants Bank

231 S. LaSalle Street
Other names: Continental Illinois Bank, Bank of America; currently
    Wintrust Financial Corporation
Completed in 1924
Architectural firm: Graham, Anderson, Probst & White
Foundation: Rock caissons
Geologic features: Salem Limestone, Chiampo Limestone, Hauteville
    Limestone

In this neighborhood, keeping up with the neighbors requires having either a neoclassical design or a Salem Limestone exterior; this building dutifully complies by supplying both. In this case, the Salem is of the "Variegated Indiana" brand. Here the architectural order is Ionic and the entrance columns are graced with capitals with characteristic large volutes.

Inside, sedate stateliness turns to splendor. In 1922, before construction began, *Stone* magazine announced that the bank project would feature "one of the largest contracts for interior marble work in recent years," awarded to Boston's Johnson Marble Company. And when the job was done, the marble-trade journal *Through the Ages* trumpeted the results: a second-floor main business area, now called the Grand Banking Hall, that boasts 28 fluted columns, all suitably Ionic, of the "Cunard Pink" variant of northern Italy's Chiampo Limestone. It was also used for the wall cladding. In contrast, the Hall's flooring is another beautiful, polishable limestone, the Hauteville from eastern France.

The Chiampo formed in the Eocene epoch and is famous in paleontological circles for sometimes containing good specimens of *Nummulites*, a genus of very large single-celled marine foraminifers that drifted as plankton in the warm and sunlit waters of the Tethys Ocean. However, the main virtue of the Chiampo's "Cunard Pink" form is its color—pink indeed, of a veined and yellowish-tannish sort. The Hauteville, of Cretaceous age, is a product of an earlier version of the Tethys marine environment. It too can show signs of ancient life, ranging from ammonite fossils to signs of bioturbation (the disturbance and reworking of the stone's original limy sediments by burrowing organisms). It ranges in color from yellow to beige or light pink.

7.12   Continental & Commercial Bank

208 S. LaSalle Street
Other names: Continental National Bank, City Bank
Lower portion now occupied by the JW Marriott Chicago Hotel
Completed in 1914
Architectural firm: D. H. Burnham & Company, then Graham,
     Anderson, Probst & White
Foundation: Rock caissons
Geologic features: Concord Granite, Terra-Cotta

This city-block-sized edifice was the first-built of the grand troika of foot-of-LaSalle fortresses of finance. It exudes the standard rock-solid-temple-of-commerce trope. In this locale, where all three of the main architectural orders have been trotted out within a stone's throw of one another, the Burnham team started the neoclassical ball rolling by choosing the simplest, Doric. The massive eastern-façade columns and the rest of the exterior's base, best examined on the LaSalle Street side, are faced in Concord Granite. An appropriately oligarchic choice, it's fine-grained, moderately gray, and devoid of dramatic contrasts. The slightly roughened, axed finish also adds to its cool demeanor. Still, at close range its bits of muscovite mica produce a subtly subversive effect by glittering here and there in the light. Quarried in the vicinity of New Hampshire's capital (pronounced KON-kud by Granite State natives), this rock type dates to the upper Devonian, a time marked by the collision of the peripatetic microcontinent Avalonia with what was then the southern margin of Laurentia, North America's Paleozoic predecessor. The result was the Acadian Orogeny, during which a great mountain chain rose and a large swarm of magma bodies was emplaced in the upper crust of what is now New England.

The Continental & Commercial's fourth through seventeenth stories seem to be clad in Salem Limestone of the same shade used by its equally august neighbor banks directly to the south. But in fact it's look-alike terra-cotta: an excellent example of cost-saving, expedient stone mimicry.

7.13   The Rookery

209 S. LaSalle Street
Original name: Central Safety Deposit Company Building
Completed in 1888
Architectural firm: Burnham & Root
Foundation: Shallow and isolated spread footings of beam and rail grillage type (the first full use thereof in Chicago)
Geologic features: Graniteville Granite, Chicago Anderson Pressed Brick, Northwestern Terra Cotta

When considered together, the Rookery and the Monadnock Building make an excellent couplet of comparisons. Both are the products of the design genius of John Wellborn Root and his fruitful collaboration with Daniel Burnham. Both embody the Richardsonian Romanesque predilection for the dark-hued edifice, even if one of them makes few other claims to that style. And both rely on the same building materials to produce the powerful sense that they're real geologic entities in their own right, great landforms rising out of the local substrate.

Here at the Rookery, as at the Monadnock, the ground-floor exterior features the "Missouri Red" variety of the Mesoproterozoic Graniteville Granite, but in this instance as stout polished columns and rugged rock-faced ashlar. The smooth-faced masonry above is the same "Brown Obsidian" variety of Chicago Anderson Pressed Brick, complemented with matching ornament crafted by another local firm, the Northwestern Terra Cotta Company. The Anderson works was located along the North Branch of the Chicago River just downstream from the Northwestern facility, at what is now N. Elston and W. Webster Avenues. It used Pennsylvanian-subperiod shale or underclay shipped in from the coal fields of the Illinois Basin in the manufacture of its top-quality facing brick.

The Rookery has always been one of the Loop's most iconic and admired buildings. In 1888 even the *American Architect*, in that era always eager to point out what it considered, from its Boston Brahmin perch, the faults of this western upstart city, had to give the Rookery substantial if not unmixed praise, proclaiming it, "all things considered, the most satisfactory of the Chicago office buildings. A great deal can be said against it, but there is so much that is good in detail, that it easily holds its place as the best designed structure of its kind." The article goes on to describe it as "built entirely of brick, a favorite material with the Chicago builders but unfortunately (we say "unfortunately" advisedly) the brick is a dark chocolate color. Had the same forms been followed in the strong cherry tones of the Insurance Exchange, which is directly opposite the Rookery, we believe the results would have been much more pleasing."

Regarding the choice of brick color, I couldn't agree less, though it's true that after Root's untimely death at age 41 Burnham largely turned his back on the somber power of clay and stone in deeper shades. Embracing the Beaux Arts ideal, he and later colleagues did their best to make the city bright and white, or at least light gray. But how fortunate we are that no one prevailing fashion or philosophy really ever succeeded in driving out the others. In Chicago both the chthonic and the rational, the explicable and the ominous, are powerfully present and patently essential.

7.14   Field Building

135 S. LaSalle Street
Other name: LaSalle Bank Building
Currently known by its street address
Completed in 1934
Architectural firm: Graham, Anderson, Probst & White
Foundation: Rock caissons
Geologic features: Mellen Gabbro, Salem Limestone, Yule Marble,
    Paitone Breccia, Nickel Silver, Aluminum

Of all the compelling recitations of the Grand Art Deco Formula to be found in Chicago, the Field Building is the most systematically imposing in its colossal symmetry and relentless vertical emphasis: one soaring central tower guarded by four smaller blocks at its corners. All are faced with Salem ("Bedford," "Indiana") Limestone adorned in its lower reaches by aluminum trim and spandrel decorations. This building, inside and out, is a cogently interwoven fugue of stone and metal.

At the base and framing the attenuated multistory entrances is a glistening black stone polished to a high-grade, looking-glass sheen. This is the "Rosetta Black" variety of Wisconsin's Mesoproterozoic Mellen Gabbro, a coarse-grained intrusive igneous rock composed mainly of plagioclase feldspar, with lesser amounts of other minerals, especially augite, a pyroxene mineral, or hornblende, depending on exactly where the rock was quarried. The Mellen was a favorite choice of Windy City architects in the 1920s and 1930s. Unfortunately, it's no

FIGURE 7.3.   In the Field Building's ground-floor corridor piers of gleaming Yule Marble from Colorado enclose book-matched panels of northern Italy's Paitone Breccia.

longer available for architectural use, even though "black granites" both foreign and domestic are once again in great demand.

The lofty piers that rise within the Mellen frame have been described in the architectural literature as everything from nickel silver to limestone to marble. The third of these is correct; it appears to be the unpolished equivalent of the gleaming Yule Marble similarly adorning the lobby piers inside. Here on the exterior you can still make out something of the Yule's crystalline texture, as well as faint veining, despite its weathered surface and occasional scrapes, scratches, and chipped edges.

Make sure you also step inside: the Field Building's Art Moderne lobby is a required stop for the urban geologist. Two rock types, the cream-colored Yule in all its polished glory and the rosy-tan to brown Paitone Breccia, constitute the cladding here. And they're magnificently complemented by an abundance of nickel silver metalwork. The Yule, a Mississippian-age true marble of exceptional quality, comes from a remote quarry locale sequestered deep in the Colorado Rockies. In contrast, the Jurassic-period Paitone hails from the northern Italian province of Brescia. This makes it, I suppose, a Breschia breccia, and one that is produced in the same neighborhood as the more frequently encountered Botticino Limestone. Here, the Paitone is represented by its especially striking "Loredo Chiaro" variety. In it you'll easily recognize the large, angular chunks of limestone that were broken apart and subsequently recemented in its calcite matrix. In this building it's been so often set in book-match fashion that the long arcade seems one great Rorschach-ink-blot gallery.

The nickel silver also deserves special mention. An alloy of copper, nickel, and often zinc as well, it in fact contains no real silver. First used in ancient China, it was in more recent times a choice highlighting element of architects in the age of Art Deco and Art Moderne. Its namesake component, nickel, is a metal familiar enough to anyone archaic enough to still carry US-currency coins. Actually, though, it's mostly concentrated in the Earth's core, where it makes up to one-fifth of that superheated spherical mass (the rest being elemental iron). As geophysicists have suggested, nickel may well be an element of supreme importance to us all, in that it plays a crucial role in the existence of the planetary magnetic field that protects us and all other surface creatures from the deadlier forms of cosmic radiation.

Because of its varying copper content nickel silver can be tricky to distinguish from competing ornamental metals. Sometimes it's produced to have truly silver color; sometimes, as here, it's something yellower, resembling pale bronze or brass. From the entrance doors and hallway trim to the counter edges and light fixtures, the Field Building's lobby is Chicago's best showplace of this elegant alloy.

7.15   190 S. LaSalle Street

Completed in 1987
Architects: John Burgee Architects with Philip Johnson
Geologic features: Porriño Granite, Alicante Limestone, Botticino
    Limestone

This is one of the city's preeminent examples of postmodernism. It vividly demonstrates this style's renunciation of the minimalism of the preceding, International school that so successfully populated the world with an endless array of glass-and-steel boxes. In the postmodern alternative, ornamental flourishes and decorative stone have returned with a vengeance. And the more exotic the stone, so it seems, the better. In the Harold Washington Library and the Olympia Centre, the fancy red granite was imported from Sweden; here it came from Spain. Known in the trade as "Spanish Pink," it's the Porriño Granite, and it's used at the base of 190 LaSalle's exterior in both polished and much rougher flamed finishes. A striking, coarse-grained igneous intrusive rock, it is quarried near the Galician town of O Porriño, just north of the Portuguese border. It solidified out of a body of magma in the lower Permian period, at the end of the great Variscan mountain-building event triggered by the collision of western Europe and Gondwana, during the formation of the supercontinent Pangaea.

Fortunately, the rock types used indoors here have also been amply documented. In the monumental, high-vaulted lobby there's another wonderful stone from Spain on display: the Alicante Limestone, which forms the salmon-colored wainscoting, pilasters, and trim. Note the white veins of calcite that crackle through it like electrical arcs. Architects consider the Alicante a marble because it's a relatively soft stone that gladly takes a polish. Jurassic in age, it hails from the opposite end of the Iberian Peninsula, just a few miles west of the Mediterranean seaport of Alicante. Above it, the wall cladding is a more restrained off-white to beige. This is another Jurassic limestone, the Botticino, from northern Italy. It has graced Chicago interiors for well over a century now, and has a history of Old World use stretching back to ancient Roman times. It can be identified by its squiggly stylolites, understated nodular mottling, and occasional fossils.

7.16   150 S. Wacker Drive

Completed in 1971
Architectural firm: Skidmore, Owings & Merrill
Geologic feature: Impala Gabbro

A stop at this site is, in effect, a visit to South Africa and a very different time in Earth history. The polished black stone that adorns the building's piers and spandrels is Impala Gabbro, quarried in the Transvaal community of Bosspruit, about 60 miles northwest of Johannesburg. It is part of what geologists call the Bushveld Complex, the world's largest layered deposit of igneous rock. This giant mass of exotic stone types and abundant rare-metal deposits was apparently the result of continental rifting and thinning of the crust at about 2.05 Ga, during the Paleoproterozoic era. Magma from all the way down in the mantle welled up through the cracks, with some of it actually reaching the surface. Though its relatively large crystals indicate that it solidified somewhat short of that, the Impala Gabbro nevertheless comes from one of the uppermost layers in the deposit. To find this messenger from the mantle here, in the heartland of another continent half a planet away, is nothing less than miraculous.

THE LOOP

Northwestern Quadrant

8.1 Hyatt Center

71 S. Wacker Drive
Completed in 2005
Architectural firm: Pei Cobb Freed & Partners
Geologic feature: Rapidan Diabase

This twenty-first-century skyscraper provides a good look at one of the "black granites" currently in favor with architects. Both the stone-clad sections of the exterior and the lobby area inside feature the "Virginia Mist" brand of the Rapidan Diabase. This igneous rock is intrusive, so its crystals are visible to the naked eye, but even so they're considerably finer than a gabbro's. Quarried between the towns of Culpeper and Rapidan, it's Jurassic in age and owes its existence to the breakup of Pangaea in the heyday of the dinosaurs. As what is now northwestern Africa broke away from the American East Coast, the continental crust experienced "extensional tectonics"—it thinned and began to rupture, with mafic magma welling up into the breach from Massachusetts to New Jersey, Pennsylvania, and the Rapidan's quarrying region in Virginia. Today this diabase is prized for its patterning: strands of light-gray plagioclase feldspar float in a matrix of darker clinopyroxene minerals like wisps of fog drifting through the night air.

8.2   Northern Trust Building

50 S. LaSalle Street
Completed in 1905
Architectural firm: Frost & Granger
Foundation: Rock caissons
Geologic feature: St. Cloud Granite

Two varieties of St. Cloud Area Granite have been documented for this building's exterior: "Rockville" and "Opalescent," in different finishes. This coarse-grained igneous intrusive rock type formed in the Paleoproterozoic era as part of the massive Minnesota Batholith. While in general it has held up well here, there are some subtle signs of spalling and disintegration where the damp course meets the sidewalk. More than likely, this is due to the infiltration of deicer salts dissolved in meltwater. As long as such compounds are used to make the pavement safe in winter—and one must argue they're essential for the greater good—this basic weathering process and the deterioration of exterior stone it causes is virtually inevitable. Breaking down and building up are both inevitable parts of the Earth's great Rock Cycle. Nature, at least on this planet, ultimately recycles everything irrespective of our preferences or willing participation.

8.3   New York Life Building

39 S. LaSalle Street
Other name: LaSalle-Monroe Building
Now known simply by its address
Completed in 1894, 1898 (eastern half), 1903 (top story)
Architects: Jenney & Mundie (1894 and 1898), architect unknown
   (1903)
Foundation: Shallow, beam grillages
Geologic features: Hallowell Granite, Northwestern Terra Cotta

This venerable LaSalle Street landmark was designed by the firm of William Le Baron Jenney, the architect who played such a pivotal role in the development of Chicago's early iron- and steel-framed skyscrapers. Geologically, though, it's a lesson in ambiguity. The sleek website of a major modern stone producer proudly proclaims the building's lower exterior is an exemplar of the use of North Jay Granite. But a much more primary source, a detailed and unequivocally worded *Inland Architect* article of 1894, makes it clear that the stone used was in fact the

Hallowell Granite. When I checked with an official of the stone company, he told me his North Jay attribution came second-hand from a previous owner of the quarry producing it. The lack of more solid documentation is troubling, but it's not proof that he's wrong, either.

The problem is that both of these intrusive igneous rock varieties are maddeningly similar in appearance. I have spent far too much time peering at this site's ground-floor cladding, and accomplished little more than the amusement of scientifically undercommitted passersby. But after much scrutiny, image-matching, and text-checking I've gradually moved into the Hallowell camp. One subtle difference between the two fine-textured, pale-gray granites is that while the North Jay's feldspar crystals usually shout out their bone-whiteness, the Hallowell's feldspars are just a trifle bluish and hence a bit more muted. Sensing a lack of substantial shouting here, I've cast my vote but remain open to correction. (Of course, it's possible that a little North Jay was subsequently added during some restoration work, and it blended in with the original all too well—in which case both sources are right.)

The Hallowell, highly regarded by late nineteenth-century Chicago builders, was also the granite with the most marketing presence in our city. Ads extolling its virtues were common in trade journals of the time. Like the North Jay, it was one of Maine's superior gray granites of Devonian age. It was also a common choice for impressive gravesite memorials.

Fortunately, the identity of the cladding of most of the New York Life exterior—everything above the third floor—is not at all controversial. It's a buff-tinted terra-cotta provided by Chicago's Northwestern works. It complements the more neutral gray of the underlying granite admirably. You'll need to cross the street and see the building as a whole to fully grasp what a felicitous match the stone and burnt clay here really is.

8.4   Inland Steel Building

30 W. Monroe Street
Completed in 1958
Architectural firm: Skidmore, Owings & Merrill
Foundation: 85-foot steel piles to bedrock (the first time used for a
    large building in Chicago)
Geologic feature: Stainless Steel

When I reach this matchless masterpiece of modernism on my tours, I show my companions its primal source: a lump of stone I've been lugging around in my knapsack. Roughly the size of a baseball but weighing seven times as much as one, my specimen is striking to the eye. It doesn't resemble any other rock type to be seen in this city. In it layers of grape-purple chert alternate with those of sparkly specular hematite, a type of iron oxide. This rock, I explain, is BIF, Banded Iron Formation also known as taconite and jaspillite. While it's not used as a decorative stone in architecture, it is our civilization's primary source of high-grade iron ore, and therefore of steel.

My chunk of BIF comes from the old open-pit Republic mine in Michigan's Upper Peninsula, but I have similar samples from the Mesabi Iron Range of northeastern Minnesota. The same basic rock type can also be found in various other Archean and Paleoproterozoic outcrops across the planet. But despite its transcendent economic importance and relative abundance, BIF remains stubbornly enigmatic. Its origins are still hotly debated. It may have formed in ancient seas much different than our own, where anaerobic bacteria converted a rich supply of dissolved ferrous iron into iron-oxide compounds that settled to the ocean bottom. On the other hand, perhaps those iron oxides were produced abiotically during episodes of intense underwater volcanic activity. Or perhaps both these factors and others, too, contributed to the widespread process of ore formation. What is clear, regardless, is that the BIF deposits are the very definition of a nonrenewable resource. Their creation required special conditions that have not existed on our planet, with a few minor exceptions, since about 1.8 Ga. Coincidentally or not, this was also the time when the oxygen content of the Earth's atmosphere was rising dramatically.

Today, the great iron ranges of the world still provide our Johnny-come-lately species with abundant opportunities to mine their deposits, crush the stone, discard the chert, and process the iron-bearing minerals into concentrated taconite pellets. These in turn are shipped to steel mills for further transmutation into a multitude of metal products. When this architectural landmark was built, much of that taconite was shipped on great ore boats from Lake Superior terminals to the steel-making centers of the lower Great Lakes littoral. I remember beholding in my childhood the Dantean vista of Gary, Indiana, of those days: the ceaseless motion of ore-hoppered trains and trucks, the choking acrid air, the sky filled with naked blast-furnace flames. Reflected in a young boy's eyes, those things were so terrifying and toxic they were beautiful.

But at the other end of that hellish production stream stands this loveliest of International Style buildings, chastely attired in its green windows and unblemished silver skin. Were it clad in regular carbon steel instead, it would be a badly corroded eyesore by now. Its stainless-steel cladding avoids that unwanted death by weathering because it's an alloy containing two rarer metals, nickel and chromium, as well. The result here is a remarkable island of chemical inertness in a brutally caustic ocean of air.

FIGURE 8.1. An enduring ode to the beauty of bare metal, the Loop's Inland Steel Building is modernist architecture at its very best. The iron that is the basis of its stainless-steel exterior formed on an ancient version of the Earth that had an atmosphere and oceans very unlike our own.

8.5   First National Bank of Chicago

10 S. Dearborn Street

Other names: 1 First National Plaza, currently Chase Tower

Completed in 1969

Architectural firms: Perkins & Will; C. F. Murphy Associates, associate
  architects

Foundation: Rock caissons

Geologic features: Town Mountain Granite, Chicago Common Brick

In the 1893 Rand, McNally guidebook to Chicago, an unjustifiably anonymous writer described the original First National Bank on this site in simple but beautiful prose: "The building offers a spectacle of handsome proportions, combining strength, durability, and great size. It is surrounded on all sides with light and air." Ironically, there's no better description of the successor structure that now occupies the whole city block. This great curving upward is indeed a spectacle, a marriage of poise and power. Taken in aggregate with the openness around it, it manages to be both gigantic and congenial. It profoundly makes sense where it is.

The building's exterior stone, and the plaza pavers too, are Town Mountain Granite, taken from the Llano Uplift of Texas, a structural feature created during the far-ranging Grenville Orogeny and the consolidation of the supercontinent Rodinia. This coarse-grained and porphyritic igneous intrusive rock formed in the uppermost Mesoproterozoic. Though their colors are dulled here somewhat by the unpolished finish, the main minerals can still be distinguished: black biotite, gray quartz, white plagioclase, and pink microcline.

The great plaza on the tower's southern side is one of America's most successful urban open spaces. On its eastern flank stands a famous contribution to the Loop's public-art collection, Marc Chagall's *The Four Seasons*. Now better protected from our unforgiving climate by a canopy, this mosaic features, in addition to all its bits of colored glass, red and salmon fragments of Chicago Common Brick. It's a subtle but fitting salute to the workaday clay product that has contributed so much to the building of this city.

8.6   3 First National Plaza

Completed in 1981
Architectural firm: Skidmore, Owings & Merrill
Geologic feature: Milbank Area Granite

This is the most accessible Loop site for examining a tried-and-true staple of the modern decorative-stone trade. The exterior cladding here is the extremely popular "Carnelian" brand of the Milbank Area Granite, which has become, for all of its interesting geology, something of an architectural cliché. When it's presented in honed or flamed form, it often appears a rather nondescript pale pink. But in polished and other smoother finishes it takes on the maroon, purplish, or reddish-brown coloration that can be characteristic of the quartz gemstone for which it's named. Milbank Area Granite, dated to the Neoarchean, is produced in a small quarrying district comprising a few rural communities that straddle the border of northeastern South Dakota and west-central

Minnesota. "Carnelian" comes specifically from Milbank in the first-named state, but it shares the same age, origin, and basic mineralogy with all the granite varieties on both sides of the border.

8.7   Chicago Loop Synagogue

16 S. Clark Street
Completed in 1957
Architectural firm: Loebl, Schlossman & Bennett
Geologic features: St. Cloud Area Granite, Bronze, Brass

Set snugly in the dwelling-place of giants, the diminutive two-story Chicago Loop Synagogue nevertheless features a façade whose stone component is the coarse-grained "Diamond Gray" variety of St. Cloud Area Granite. This Paleo-proterozoic intrusive igneous rock, quarried near Cold Spring, Minnesota, is here presented in a sawed-finish form that offers a more understated aspect than its polished equivalent would. The large sculpture *Hands of Peace* by Henri Aziz is mounted on the northern side. Reportedly made of both bronze and brass, it is all the more striking for its weathering patina of contrasting black and blue-green that signals the copper content of these two alloys.

8.8   Avondale Center

20 N. Clark Street
Completed in 1980
Architectural firm: A. Epstein & Sons
Geologic feature: Aswan Granite

In the case of ornamental stone, as in much else, the concept of longevity is a relative one that can be approached from different perspectives. While the Morton Gneiss displayed on various Chicago buildings is unquestionably the oldest rock type to be found in the city, the Avondale Center's richly red Aswan Granite, here used in a nonreflective flamed finish, has in its own way an equally impressive claim to antiquity. For it boasts the longest record of human use, which extends back over 5 ka. In contrast, the Morton has been available to the architect for just a single century.

So the handsome igneous rock that adorns this otherwise nondescript building is of special interest to the archaeologist. But its unusual mineralogical makeup also tickles the curiosity of the geologist. The rock is composed of some of the usual ingredients—black biotite, gray quartz, and pink to brick-red potassium feldspars. Still, something important is missing. It's the plagioclase feldspar, normally a major constituent. This omission may seem trivial—after all, this rock certainly *does* look like a granite—but to the igneous petrologist, it's a flashing red warning light that indicates the Aswan Red is a relatively rare A-type granite. This means that unlike most of the other rocks of its kind used in Chicago, this one did not originate in a mountain-building event on a continental margin, where one plate was subducting under another. Instead, its parent body of magma formed in a continent's interior, and at unusually great depth—probably just above the crust-mantle boundary. This in turn may mean that it was associated with a mantle plume, a great column of very hot material rising from even deeper in the Earth's interior. When they burn their way up to the surface, mantle plumes create areas of intense volcanic activity called, with considerable understatement, hot spots. Modern examples include the Hawaiian Islands and Wyoming's Yellowstone caldera. In the case of the Aswan Red, this happened late in the Neoproterozoic era, about 600 Ma.

To bring the Aswan Red back into the much tighter scale of human history, imagine that you're one of this granite's earliest producers, living in ancient

FIGURE 8.2. A typically inquisitive urban geologist conducts a highly sophisticated digital analysis of the Avondale Center's Aswan Granite, the building stone with the longest record of human use found in Chicago.

Egypt's Early Dynastic Period, in 3,000 BCE, some fifty centuries ago. You work in the sun-seared landscape of the eastern bank of the world's longest river, near what one day will be the modern city of Aswan. That's 500 miles upstream, as the ibis flies, from where the northward-flowing Nile debouches into the Mediterranean Sea. The technology you use to shape the granite for your culture's architectural and artistic uses falls a trifle short of modern methodology and occupational-safety standards. While some rudimentary, soft-bladed copper saws may be on hand, and while loose sand can always be used for abrading and polishing, your main tool is just another piece of rock. It has to be light enough for you to hammer with for long hours on end, but also heavy and hard enough to crack the granite, at least eventually. The diabase that outcrops locally will do. However, you don't waste your time whacking away at a solid headwall of granite; you take your chunk of diabase and pound it instead, again and again, on one of the large boulders of Aswan Red that have already become detached through the process of natural weathering. All in all, this is not a preferred career path for anyone with a short attention span or the need for immediate gratification.

Such reliance on raw human muscle power remained the sole method of production for many centuries thereafter. From our distant vantage point, we can only imagine the flurry of excitement that erupted when, almost three millennia later, the use of iron wedges finally became a common practice in Aswan quarrying. But it's easy to picture the dismay of the industry's old-timers, who probably shook their heads in disgust and predicted that this new app would never catch on.

8.9   St. Peter's Church and Friary

110 W. Madison Street
Completed in 1953
Architectural firm: K. M. Vitzhum & J. J. Burns
Geologic features: Murphy Marble, Milford Granite

"Nothing is more strikingly indicative of the material spirit of the age than the fact that every house of God has been driven from the main business streets of Chicago." This view, expressed in the 1893 Rand, McNally guide to the Windy City, no longer pertains. Along with the nearby Chicago Loop Synagogue, St. Peter's has ministered to the Loop's faithful for the past seven decades. It's an interesting urban adaptation, a pocket house of worship in this skyscrapered vertical landscape. And how appropriate that it bears the name of the saint Christ identified as the rock upon which his church would be built.

The rock that makes the façade of this building so visually arresting is the pink and gray-striped "Etowah" variant of the Murphy Marble. Here amid the dappled light and broad shadows of the urban canyon it takes on an unsettled and almost chaotic aspect, swirling about the anchoring solidity of the massive central crucifix. The "Etowah" is one of a family of superb Murphy Marble selections produced in the vicinity of Tate, Georgia. As with other metamorphic rocks, it's best to speak not of its age, but of its ages: first of the origin of its parent rock, then of its transformation into its current form. In the case of the Murphy, it started as a limestone, probably in the Cambrian period, and at some later time or times in the Paleozoic era it was subjected to enough heat and compression to recrystallize its calcite and produce the rich waves and billows of gray graphite.

Beneath the Murphy there runs a damp course of Neoproterozoic Milford Granite from southeastern Massachusetts. It is indeed pink by granite standards, but with its considerable black biotite-mica content it provides a distinct change in both tint and texture. Its other characteristic minerals are white albite, light-gray or bluish quartz, and pale-pink orthoclase and microcline.

8.10   1 N. LaSalle Street

Completed in 1930
Architectural firm: K. M. Vitzhum & J. J. Burns
Foundation: Rock caissons
Geologic features: Salem Limestone, St. Cloud Area Granite, Brass

Another exemplar of the Grand Art Deco Formula, this building is an elegant rectilinear mountain composed of set-back blocks of uniformity-enforcing Salem Limestone, all resting on a base of crystalline rock. What makes this building so distinctive from street level is its elaborate decorative patterns in the Salem, both in the framing of the monumental entrance and up on the fifth floor, where there is an entire series of panels depicting the adventures of René-Robert Cavelier, Sieur de La Salle, who, in the process of exploring and claiming the great river systems of central North America for King Louis XIV of France, supposedly encamped on this very spot in 1679. All this ornament, the work of French-American sculptor Leon Hermant, demonstrates the amazing ability of this limestone to take intricate carving that still retains its details many decades later.

Also note (and it's pretty difficult not to) the magnificent entryway with its gleaming decorative panels of brass, the alloy of copper and zinc that usually has a yellow-gold color that is lighter and brighter than bronze. In a city rich in metal artistry, this is some of the best. And it beautifully illustrates how in Art Deco metallic alloys and elements achieved their happiest pairings with stone.

While the medium-gray cladding stone that constitutes the bulk of the base is not documented, the very deep-gray polished panels flanking the main doorway are. They're the "Charcoal Black" brand of the ever-popular St. Cloud Area Granite family. This unusually dark variety contains equal amounts of gray plagioclase and orthoclase feldspars, as well as black hornblende and clear quartz. Paleoproterozoic in age, it's quarried about 3 miles south of the town of St. Cloud, Minnesota, in a locale that produces gray-toned rock rather than the pink and red granites found elsewhere in the region.

8.11   Roanoke Building

11 S. LaSalle Street
Original name: Lumber Exchange Building
Completed in 1915, 1922 (5-story addition), 1926 (tower)
Architectural firms: Holabird & Roche (1915 and 1922), Rebori,
    Wentworth, Dewey & McCormick (1926)
Foundation: Rock caissons
Geologic features: Midland Terra Cotta, unknown Ophicalcite or
    Serpentinite

The Roanoke Building's current ground-story exterior cladding was added during a major renovation in 1984. It acts like a powerful electromagnet on any passing rockhound or geologist. While most building stone imparts the suggestion of solidity or at most graceful, directed motion, these panels of darkest-green fragments jumbled in a web of palest-blue clots and veins instead suggest dissolution and chaos. I'm not quite sure how it's meant to relate to the sedate brown terra-cotta above it, which was crafted in Portuguese Gothic, a style rarely beheld in these parts. But the powerful pull of this insurrectionary stone makes me hardly care. This is an excellent outcropping of wild and curiosity-provoking stuff.

The exact identity of this splashy rock selection has not revealed itself. It might be classified as a serpentinite, that weirdly wonderful metamorphic stone derived from lower-crust and upper-mantle material scraped up in continental collision and then subjected to further tectonic upheaval. But more likely it's an ophicalcite, the sedimentary or sometimes remetamorphosed breccia composed of angular chunks of serpentinite. The exact demarcation line between the two rock types can be very hard to draw with any consistency, but here the amount of fragmentation of the green serpentinite clasts, and the abundance of cementing calcite (or another carbonate mineral) makes the ophicalcite tag stick well indeed.

While this stone may be a hard act to follow, the terra-cotta mentioned above is also worthy of note because it's a well-documented example of the high-quality work of the Midland Terra Cotta Company. This firm was a competitor of the better-known Northwestern Terra Cotta Works, and like it, was Chicago-based. It was acquired by the third great regional producer, Crystal Lake's American Terra Cotta Company, in 1938.

8.12   Madison Plaza

200 W. Madison Street
Completed in 1983
Architectural firm: Skidmore, Owings and Merrill
Geologic feature: Carrara Marble

On the wall of this office tower's ground-floor lobby is a magnificent exposure of the stoutly veined "Bettogli" variant of Carrara Marble. Quarried in the Apuan Alps of Tuscany, Italy, the Carrara in this and its various other offerings began as Jurassic limestones that were metamorphosed in the Oligocene epoch, during mountain building that resulted from the ongoing interaction of the European and African plates.

8.13   33 N. LaSalle Street

Completed in 1929
Former names: Foreman National Bank, American National Bank
Architectural firm: Graham, Anderson, Probst & White
Foundation: Both hardpan and rock caissons
Geologic features: Marcy Metanorthosite, Salem Limestone

This section could bear the byline "Art Deco with a Difference," because while its subject generally follows the Grand Formula in being a Salem Limestone mountain with setbacks and a crystalline-rock base, it also features some quirky variations. Among them, light- rather than dark-toned basal cladding and overtly botanical ornamentation that inclines much more toward the vocabulary of Art Nouveau.

The geological treasure here is the stone on the lowest stories of the exterior. Most of it has an unusual palest-blue tint with black flecking that looks like splatters from a leaking fountain pen. This is a striking variant of the Mesopro-terozoic Marcy Metanorthosite quarried in the Adirondacks at Au Sable Forks, New York. Formerly marketed as "Ausable Blue" and more recently as "Lake Placid Blue," it here proves itself to be a cladding material that is an especially attractive and enduring medium for fine ornamental carving. Take a good look at the manic mixture of leaf, stem, and flower over the southern doorway.

There's another rock type here as well, a much more foliated and contorted green stone set in tall panels on either side of the main entrance. While documentation for the Marcy Metanorthosite has surfaced, records of this selection's identity sadly have not. But its tight, ptygmatic folds suggest that it's a gneiss.

8.14   Conway Building

111 W. Washington Street
Completed in 1913
Current name: Burnham Center
Architectural firm: D. H. Burnham & Company
Foundation: Rock caissons
Geologic features: Hinsdale Granite, Concord Granite, Vinalhaven Granite

This Burnham Beaux Arts beauty has on its ground-floor exterior an assortment of granites of different hues. One of these is the "Duluth Gray," which seems to be an old trade name for the rare Hinsdale Granite elsewhere documented only for Adler and Sullivan's Auditorium Building. Here on the Conway Building it was identified by the anonymous 1960s Coldspring compiler for the two col-umns that stand at the Washington Street entrance. This is plausible. One of the hallmarks of that ancient and rarely seen Neoarchean rock from northeastern Minnesota is its large clumps of hornblende crystals. There are many of these black "knots" on the Auditorium exterior. Here they're fewer and smaller, but still detectable on the column shafts.

The bulk of the exterior walls is dressed in an old but distinguished depend-able, the Devonian-period Concord Granite from New Hampshire. It's a gray, fine-grained selection often identifiable by its small but very reflective bits of muscovite mica. But the low damp course under the Concord is the con-trasting pink and coarse-grained Vinalhaven Granite, of Silurian age, that was quarried on the Maine seacoast. It has often been marketed as "Fox Isle" or "Fox Island Granite" instead. It suite of minerals includes pinkish orthoclase, gray quartz, white oligoclase, and black biotite. The identity of yet another granite—the reddish one in the column bases and backing—was apparently not documented.

8.15   City Hall and County Building

121 N. LaSalle Street (City), 118 N. Clark Street (County)
Completed in 1911
Architectural firm: Holabird & Roche
Foundation: Rock caissons
Geologic features: Woodbury Granite, Botticino Limestone, Northwest-
   ern Terra Cotta

The extant City Hall and County Building replaces an earlier version, completed in 1885, that had a troubled history and a dreary record as whipping-boy of the Boston-based *American Architect*. That journal's Brahmin critics, who wasted very few opportunities to put this parvenu city of the unpolished West in its place, reveled in the fact that the exterior of the County side was Lemont-Joliet Dolostone, while the City portion, erected later, was Indiana's Salem Limestone. Admittedly, this was not a happy combination of materials. The former has a coarser, less compliant texture and weathers in various interesting and unexpected ways from bluish-gray to golden yellow; while the latter is an ultra-uniform grayish buff that, besides attracting its share of nineteenth-century Chicago's ubiquitous soot, changes not at all.

In this description I focus on the eastern, County side of the current Corinthian-columned complex composed of two massive, mirror-image structures that happily are clad in the same stone type. In 1907 the *Western Architect* reported that this section alone rests on 130 caissons, 4 to 10 feet in diameter, that extend downward 110 to 120 feet below street level and contain 450,000 cubic feet of Portland Cement.

The Washington Street façade across from the Conway Building is ornamented with bronze-colored trim and details produced by the Northwestern Terra Cotta Company. But the predominant feature is the cladding of Woodbury Granite, quarried on Robeson Mountain in the northern Vermont community of Woodbury. The entire City/County complex wears about 500,000 square feet and 42,500 tons of it. This medium-grained, salt-and-pepper stone has a certain undefinable monumental aspect to it that is particularly well suited to neoclassical government and business headquarters. Its suite of minerals includes three lighter-tinted feldspars—microcline, orthoclase, and oligoclase—as well as smoky quartz and black biotite mica. In common with New England's other Devonian granites it owes its origin to the Acadian mountain-building event that in turn resulted from the collision of the microcontinent Avalonia with what is now the Northeast.

Once you're inside the County Building, memories of the cool granitic gray of its exterior quickly give way to the corresponding coolness of the beige and off-white of the hallway walls and piers. This is the Botticino Limestone, a subtly mottled Jurassic carbonate rock. It's the kind of polishable sedimentary stone architects call marble. An import from the Alpine foothills of northern Italy, it has a two-millennium legacy of use in the Old World. And now, in Chicago, it's one of the more frequently encountered interior-cladding selections in grander settings.

8.16   Richard J. Daley Center

66 W. Washington Street
Completed in 1965
Original name: Chicago Civic Center
Architectural firms: C. F. Murphy & Associates; Loebl, Schlossman & Bennett; Skidmore, Owings & Merrill, associate architects
Foundation: Rock and hardpan caissons
Geologic features: Weathering Steel, St. Cloud Area Granite

Certainly one of the most striking of the city's extensive roster of International Style skyscrapers, the Daley Center is also one of the easiest to take in at a glance because of its open plaza that gives it the monumental setting it deserves. The building rests on 50 caissons, which, at 120 feet deep, make contact with the Silurian dolostone bedrock. The plaza is supported by no less than 108 caissons sunk to hardpan at 86 feet.

Both the Daley Center's exterior and the famous Picasso sculpture in front feature Cor-Ten, the variety of weathering steel first developed in the 1930s by the United States Steel Corporation. Originally intended for use in ore-carrying railroad hopper cars, it was not employed architecturally until the 1964 construction of the John Deere corporate headquarters in Moline, Illinois.

Weathering steel in its various brands and alloys has the virtue of quickly forming an oxidized outer layer that dramatically slows subsequent corrosion. In essence, it rusts to prevent more rusting. This is what scientists refer to as a negative feedback loop, an initial condition or effect that triggers its own diminishment. The characteristic color this process produces can invite immediate associations of disrepair and decay: the car hulk rotting in a rural front yard, the group of old, weathered 55-gallon drums suspiciously leaking some foul-smelling fluid in an abandoned lot. But in fact the color of rust is one of nature's primal hues, and, once one gets used to it, it's compellingly attractive. After all, we live on a planet with plenty of iron in its crust and an atmosphere with an anomalously high oxygen content ceaselessly replenished by the activity of photosynthetic organisms. They, more than anything else, have made the Earth one rusty mother. Magnificent red- and sunset-tinted landscapes in the American Southwest and elsewhere show how much iron oxide is now on display in the rock record. Accordingly, one could make a cogent case for the Daley Center, this magnificent architectural ode to corrosion, being one of the most overtly geological buildings in the city and one of the most illustrative of our world's basic processes. The iron and other metals present in the steel were first extracted from the crust—a sort of high-speed, anthropogenic erosion—and then subjected to the same inexorable force of weathering that affects every surface exposed to wind and water. But in this case, our species has learned how to make that process both self-limiting and beautiful.

While the weathering steel is this site's main attraction, don't neglect to notice the plaza's pavers, which are a variety of the Paleoproterozoic St. Cloud Area Granite quarried in Rockville, Minnesota. This intrusive igneous rock, which has been used extensively throughout the metropolitan region, was part of the East-Central Minnesota Batholith that solidified in the upper crust about 1.78 Ga ago. The same handsomely coarse-grained stone is on display in the building's lobby, as pavers and cladding for the elevator banks.

8.17 Commonwealth Edison Substation

115 N. Dearborn Street
Completed in 1931
Architectural firm: Holabird & Root
Geologic features: Aswan Granodiorite, Salem Limestone

All too often overlooked amid the clutter of the subsequent Block 37 "development," this Art Deco gem deserves more reverence than it currently receives. The surrounding gaggle of glassy giants is dwarfed, architecturally speaking, by this smaller elder sibling. At time of writing its blocked-up Dearborn entranceway and the rest of its lower stone exterior is in serious need of a good cleaning.

A 1930 *Chicago Tribune* article describing the impending construction of this utilities facility noted that "it's to be faced with huge slabs of Indiana limestone resting on a base of Egyptian black granite." So once again we have a recitation of the Grand Art Deco Formula, but this time in miniature: the usual variety of buff-colored sedimentary carbonate above and something darker, crystalline, and more exotic below, all stuffed into a mere four stories. The upper limestone, geologically dubbed the Salem, was the standard facing material of the era, much appreciated for its durability and willingness to take and retain intricate carved detail. Here it bears a bas-relief figure, *The Spirit of Electricity*, designed by Sylvia Shaw Judson.

The real prize, though, is the "black granite" below and in the shut-up door-way. In the assumption that it has not been entirely replaced by some remark-ably similar substitute, this matches in appearance, texture, and composition the Aswan Granodiorite, still quarried, as it has been for thousands of years, on the eastern side of the Nile in southern Egypt. To my knowledge, this is the only documented use of this archaeologically famous stone in Chicago's built landscape, and it can rightfully be considered one of the city's rarest. It shouldn't be confused with the pink to red Aswan Granite on display just two blocks away on the Avondale Center. What distinguishes granodiorite from true granite is a technical point of mineralogy—its feldspar content is mostly in the form of plagioclase, whereas granite has a more equal mixture of the plagioclase and alkali feldspars.

Like its companion red granite, the Aswan Granodiorite is late Neoproterozoic in age. The ancient Egyptians, who were rather sophisticated and detailed rock classifiers, specifically referred to this one stone variety as *inr km* (apparently pronounced "iner kem"). Its somber cast is due to the predominance of dark-gray quartz and black biotite and hornblende. In some specimens these are joined by large pink or white feldspar phenocrysts, but at this site they seem to be lack-ing, so the rock has a much more uniform texture. The presence of this historic and thoroughly uncommon stone type singlehandedly makes the Substation, already significant architecturally, one of our most geologically important urban outcrops.

8.18 Delaware Building

36 W. Randolph Street
Former names: Bryant Building, Real Estate Board Building
Completed in 1874, two-story addition in 1889
Architectural firms: Wheelock & Thomas (1874), Holabird & Roche (1889)
Foundation: Shallow and spread
Geologic features: Cast Iron, Pressed Metal, Concrete

The Italianate-style Delaware Building stands as a noble survivor of a once much more extensive roster of Loop buildings constructed in the first few years after the Great Fire of 1871. Take a moment to imagine this neighborhood when this now modestly proportioned edifice was erected as the latest thing in Windy City architecture. Erase all the high-rises and glass-and-steel boxes nearby; replace them in your mind's eye with buildings very much like the Delaware in height and style. Then throw caution completely to the winds and imagine the noise and press of the crowds, buggies, trolleys, and lumber wagons clogging the intersections. And dare to breathe in the air of a city powered by soft Illinois Basin coal and horseflesh on the hoof. It smells of soot, sweat, fresh dung, and rapid evolution.

The Delaware's façade is an interesting layer cake of different materials. Cast iron, already a longtime favorite for Chicago building exteriors when this structure was built, bedecks the two bottom stories. But for the four directly above them, prefabricated concrete, making an early appearance in this role and marketed as "Cast Stone," was used instead. And the top two floors, added a decade and a half later, were clad in pressed metal.

8.19   333 W. Wacker Drive

Completed in 1983
Architectural firm: Kohn Pedersen Fox; Perkins & Will, associate architects
Geologic feature: Vermont Serpentinite

Touted for its convex, sky-reflecting wall of glass that faces the Y intersection of the South, North, and Main Branches of the Chicago River, 333 W. Wacker sports a triad of stone types on its exterior. I here pass over the fact that two of them, gray and black and both described as "granites," remain unidentified because my requests for information from the building's architects and management company were met with a serene and unruffled indifference. But the most distinctive rock type of the lot is the well-provenanced, forest-green and white-veined serpentinite that simply steals the show. This striking selection, often misidentified in the architectural literature as a marble, has a trade name that's an amusing troika of English, Italian, and French: "Vermont Verde Antique." It may be a linguistic chimera and a mouthful to say, but the rock type it identifies has been popular as a decorative stone for well over a century now, in both indoor and outdoor settings. That last fact is in itself remarkable because most serpentinites degrade badly when exposed to the elements. Perhaps this variety's greater resilience is due to the fact that its veins are mostly composed of the mineral magnesite, which is less reactive to acidic rain and atmospheric compounds than the calcite more commonly found in other rocks of its type.

The Vermont Serpentinite is most accessible here in the polygonal columns at the building's front, and in the lowest courses of façade panels that alternate with those of the unidentified gray granite. This unusual stone began its existence in the Neoproterozoic or Cambrian as dunite, an ultramafic igneous rock that forms in the Earth's upper mantle. Later, in the Ordovician, it was plucked up by a migrating island-arc complex, rather like a section of topsoil being scraped up by a bulldozer blade, and shoved onto the eastern margin of North America, as part of the formation of the Taconic Mountains. Still later, in the Devonian, another continental collision created the Acadian Mountains and further metamorphosed this rock. So it's been quite a journey. From mantle to crust; from the hills of New England to the bank of an inland prairie river; from ancient interior darkness to the sunlit surface of a splashy postmodern skyscraper. Exotic in appearance, exotic in its origin, and exotic even in its triple-tongued trade name, the "Vermont Verde Antique" is the premium showboat stone in a city of showboat architecture.

8.20   LaSalle-Wacker Building

221 N. LaSalle Street
Completed in 1930
Architectural firm: Holabird & Root
Foundation: Rock caissons
Geologic features: Addison Gabbro, Salem Limestone

To this day, I can't help but see this riverside representative of Chicago's Art Deco efflorescence the way it looked in my childhood, before it was hemmed in by larger but lesser buildings. This because in the late 1950s and early 1960s my father based his small advertising agency here, and often brought me along to spend the day with him during my school breaks. I recall that this made eminent sense to me; the Loop had obviously been constructed as a sort of gigantic toy shop for my personal investigation and delight. While the boss met with clients or finished his latest project, junior was allowed to wander the building alone and explore its interior. That was in those glorious days before the onset of helicopter parenting and banks of Orwellian security cameras monitored by cadres of blazered, walkie-talkied concierges. I quickly established a meaningful working arrangement with what I gather was the building's one watchman, and under his none too watchful eye conducted sophisticated architectural surveys, primarily by running up and down the stairwell, from upper basement to the forty-first story observation deck, as many times as I could before my senior business partner reeled me in and took me to lunch at the corner coffee shop. To crib from Wordsworth, bliss was it in that dawn to be alive.

Like other members of its tribe in Chicago, the LaSalle-Wacker is mostly clad in buff-colored Salem Limestone—the "Bedford Stone" of architectural parlance—with darker crystalline-rock cladding at the bottom. Here, however, that second element is understated, and relegated to a simple damp course, higher between the display windows, and entrance trim. Nevertheless, it's a geological find of the

FIGURE 8.3. River-Fronting Loop skyscrapers as seen from across the Main Branch of the Chicago River at the Merchandise Mart (far left). The set-back building with the antenna at top is the Art Deco LaSalle-Wacker Building; the pedimented, postmodernist 77 W. Wacker is two buildings to its left. Between them stands an excrescence of The-Aliens-Have-Landed School of dystopic architecture. There goes the neighborhood.

first order. The "black granite" used here is not just black, but black peppered with white speckles: it's the Addison Gabbro of Devonian age from the rocky seacoast of Maine. It was marketed under several trade names, including "Lang's Black," but its only other documented occurrence in Chicago is on the Palmolive Building. The white mineral giving it its unusual spotted look on both buildings is andesine, a plagioclase feldspar rarely seen elsewhere in decorative stone. It's complemented by the brownish pyroxene called diallage, and by black biotite mica and small amounts of dark hypersthene and olivine.

One happy consequence of the LaSalle-Wacker's having a low damp course is that the Salem Limestone, often far out of reach on skyscrapers of this era, is here readily examinable. You'll soon locate some fine bits of Mississippian marine fossils embedded within it. The most common type in the stone is *Globoendothyra baileyi*, a type of planktonic organism belonging to the Foraminifera, or "forams." This tiny creature, about half a millimeter long, encased its single cell in a small shell that may be seen in good light with a powerful hand lens, but is more easily recognized in the lab under a dissecting microscope. Less common but much larger and easier to spot here on the street are sections of segmented crinoid ("sea lily") stems and latticelike fragments of "moss animal" bryozoan colonies. It's discoveries like these, overtly dripping with magic—rare and beautiful rocks, and signs of ancient life in unexpected places—that have reinforced rather than revised my childhood view of what this city really is.

8.21   77 W. Wacker Drive

Former name: R. R. Donnelley Center
Completed in 1992
Architects: Ricardo Bofill; DeStefano & Goettsch, associate architects
Geologic feature: Pedras Salgadas Granite

An arresting postmodernist trelliswork of stone, glass, pilasters, and pediments, 77 W. Wacker is clad in a stone various architectural sources call "white Portuguese Royal granite." I gather this is a more-or-less straight translation of the rock type quarried in Portugal and sometimes marketed as "Branco Real." And it does appear to be a perfect match for it in color, texture, and mineral content. Geologists call it the Pedras Salgadas Granite instead, after its locale of origin in the northern section of that Iberian nation.

The Pedras Salgadas Granite on view here is generally medium-grained and somewhat porphyritic, in that it has white crystals larger than the rest. Its mineral makeup includes potassium and plagioclase feldspars as well as quartz and biotite; the last provides the black flecking. This intrusive igneous rock type was emplaced as one of a swarm of plutons that rose from the depths in the final phase of Europe's Variscan Orogeny. That mountain-building event took place late in the Paleozoic era, during the assembly of the supercontinent of Pangaea. This handsome and exotic granite, all too rarely seen in Chicago, has through an analysis of its rubidium and strontium isotopes yielded a radiometric date of 297±14 Ma. This means that it most probably solidified from magma in the lowermost Permian, but could possibly be as old as upper Carboniferous instead. In any event, it's a welcome addition to Chicago's roster of beautiful rock types from around the world.

8.22   Clark Street Bridge Houses

Clark Street at the Chicago River
Architects: Loran Gayton and Paul Shioler
Completed in 1929
Geologic feature: Concrete

The mansard-roofed bridge houses of this double-leaf bascule bridge share the Beaux Arts style with those of the Michigan Avenue Bridge. At a distance they also seem to feature the same exterior material, Salem Limestone. However, they're actually concrete structures both outside and in, where that most popular of modern building materials is strengthened with rebars (reinforcing bars). They amply demonstrate that concrete, in effect a manmade limestone, can hold its own as a decorative as well as structural element. But how can it be distinguished from the Salem? Both share the same basic light-gray to buff coloration and can be similar in texture; and both fizz in dilute hydrochloric acid thanks to their calcium-carbonate content. The proof lies in what the concrete doesn't have: the marine-invertebrate fossil fragments, sometimes overtly macroscopic but sometimes only convincingly identified with a hand lens, that are the Salem's hallmark trait.

8.23   Dearborn Street Bridge House

Dearborn Street at the Chicago River
Completed in 1963
Foundation: Rock caissons
Geologic feature: Milbank Area Granite

The solitary bridge house on this double-leaf bascule bridge is well worth a close look because of its cladding. While at least a couple of its companions on nearby Chicago River spans are Beaux Arts structures decked out in Salem Limestone from the Mississippian strata of southern Indiana's Bedford region, this more rectilinear and utilitarian structure, which is about as far from the Beaux Arts look as is conceivable, is covered in the "Bellingham" variety of Neoarchean Milbank Area Granite quarried on the Minnesota-South Dakota border. This means that the rock type you see here is about seven and a half times older than the Salem and, when seen close at hand, about seven and a half times more beautiful. As is often the case with Milbank granites, it is somewhat gneissic in texture, with its black biotite and lighter orthoclase, microcline, and quartz crystals exhibiting a good deal of linear banding suggestive of metamorphic foliation.

NEAR WEST SIDE, GARFIELD PARK, AND HUMBOLDT PARK

9.1 Daily News Building

2 N. Riverside Plaza
Now known simply by its address
Completed in 1929
Architectural firm: Holabird & Root
Foundation: Both hardpan and rock caissons
Geologic features: Salem Limestone, Morton Gneiss

It's a grim thought, but if an urban geologist were ever to suffer the tragic fate of being banished from the Loop proper, he or she could still get an excellent sense of one of the city's chief architectural glories, the design ethic I call the Grand Art Deco Formula, by visiting this building just across the South Branch of the Chicago River. But to get the best sense of what a wonderful building this is, our exile would still need to sneak Loopward at least halfway across the Madison Street bridge to really comprehend the bulk and grandeur of this giant, throne-like structure.

If the Daily News Building conjures up strong memories of that other river-fronting masterpiece, 333 N. Michigan, it should. These two buildings are fraternal twins. They were born of the same architectural firm headed by John Holabird and John Wellborn Root Jr. at about the same time; and though both have their own distinctive traits they share many of the same familial characteristics.

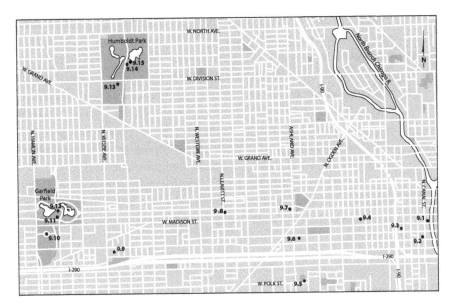

MAP 9.1.   Chicago's West Side.

Chief among these are their massive vertical exposures of that carbonate-bank biocalcarenite, the buff, Mississippian-age Salem Limestone, and the same basal cladding of Paleoarchean migmatite, the rampantly patterned, endlessly interesting Morton Gneiss. The latter here adorns the Madison Street and riverfront façades.

In those far-distant days of yore when most Americans derived their vision of the world from inked and folded sheets of processed cellulose, this imposing edifice was the headquarters of one of Chicago's great newspapers. It was also honored, and still should be, for being the first Windy City skyscraper to feature a large and accessible public open space on its lot. What is less well known, however, is that its subsurface structure is a veritable forest of caissons: 100 of which, located under its loftiest section, descend all the way to bedrock 90 feet down, while 59 more reach down just 60 feet to the hardpan, to support the building's lower flanks and plaza. All of these had to accommodate the fact that this was the first Chicago building erected over a network of railroad tracks—a situation that required especially clever engineering solutions. These included dampening the vibrations of arriving and departing trains by resetting the tracks and platforms on a concrete pad underlain by a bed of clay, and by placing asphalt gaskets under the supporting columns.

But of course it's the visible aboveground bulk of stone and steel that most directly amazes. Holabird proudly described his masterpiece in a 1929 article in the *Western Architect*:

> Unhampered by past architectural styles, a strict adherence to the solution of practical problems, expressed in continuity of line, sharp contrasts in light and shadow created through definite angular moldings and broken panes accented by ornament only at focal points, produces a definite rhythm and the definite emotion of inevitability, which is perhaps the acid test of creative endeavor along any line.
>
> The Daily News building, placidly located on the edge of a broad river, possesses this sense of inevitability to a marked degree and it is accompanied by a sphinx-like quality of feeling carried almost to the point of mysticism.

9.2   Union Station

210 S. Canal Street
Completed in 1925
Architectural firm: Graham, Anderson, Probst & White
Foundation: Hardpan caissons
Geologic features: Salem Limestone, Tivoli Travertine, Holston Limestone

The surviving portion of Union Station, the Headhouse, offers the visiting rockhound an extended essay in a trio of carbonate rocks. On its exterior, including the solemn Tuscan-order columns that line the eastern and western sides, American architecture's favorite limestone, the Salem or "Bedford," holds sway. There's so much of this famous Hoosier export in Chicago that the city could plausibly be renamed New Indiana. Inside the terminal, though, there are two other limestones with a long history of architectural use. The first, much younger than the Mississippian-age Salem, is the Quaternary-period Tivoli Travertine, which adorns the walls, columns, and staircase steps of the great Waiting Room, our region's most effective and impressive expression of the neoclassical style. This is the best place to behold this venerable rock type in its most proper setting and to understand its ornamental appeal. It was to ancient Roman builders what the Salem has been in the United States much more recently: the material of choice due to its excellent workability and abundant supply. Here, in 2015, after decades of wear and tear inflicted by untold thousands of commuting feet, the steps leading up to the Canal Street entrance were meticulously replaced with new slabs of the same stone quarried east of Rome.

FIGURE 9.1.   Chicago's finest neoclassical interior space, Union Station's Waiting Room imparts a sense of calm imperial grandeur abetted by the ancient tones and textures of the Tivoli Travertine of its walls, column shafts, and staircases. The pinkish-brown flooring of stylolitic Holston Limestone provides a gentle contrast.

The other sedimentary carbonate rock here is more easily overlooked, simply because it constitutes the Waiting Room's main flooring. Regrettably, we all have a tendency to ignore what's beneath us. But it definitely deserves scrutiny. It's the pinkish-beige version of "Tennessee Marble," known to geologists as the Holston Limestone. A frequent but always fascinating sight in the interiors of Chicago buildings, it's best identified by its stylolites, the squiggly dark lines that meander across its surface. Middle Ordovician in age, it's the grand old rock type here, at more than 100 Ma older than the Salem and 460 Ma older than the Tivoli.

9.3   Old St. Patrick's Church

140 S. Desplaines Street
Completed in 1856
Architectural firm: Carter & Bauer
Geologic features: Lemont-Joliet Dolostone, Cream City Brick

Chicago's longest-surviving church, Old St. Pat's (as its Catholic parishioners fondly call it) offers a rare glimpse of the geologically derived materials builders used in the decades before the 1871 Great Fire. Its base consists of ashlar blocks of Lemont-Joliet Dolostone quarried in the Lower Des Plaines River Valley, within easy reach of what was then the region's inland superhighway of goods and commerce, the Illinois and Michigan Canal. This handsome stone, a gift of the shallow subtropical Silurian sea that covered this region 425 Ma, is bone-white to bluish-gray when freshly cut. But it soon weathers to warmer tones—often a buttery yellow to somewhat browner color as its ferrous-iron content oxidizes and hydrates on exposure to the elements. This progression of coloration is especially visible in the ashlar here, where blocks that have suffered recent flaking and scaling reveal the original light-blue tint quite vividly.

This variation in hue gives the rock a great deal more character and ornamental appeal than the amazingly nonreactive Salem Limestone, but when the latter was coming into full fashion in the 1880s, opinions trended decidedly in the opposite direction. William B. Lord, writing at that time in the *Inland Architect*, decried the dolostone's coloration as an overt defect, and also deplored its "shaling"—a reference, apparently, to the *scaling* cited above. This is a defect, to be sure, but one shared with many other stone types, particularly when they're not well seasoned at the quarry, or not mounted properly, or not used in the appropriate places. For example, the mighty Salem has a wicked tendency to exfoliate when it's allowed to come into contact with the damp ground and winter deicer salts. It seems as though Mr. Lord was deploring nature itself, and its insistence on ultimately breaking down and recycling everything. And in any case, early Chicago buildings that feature Lemont-Joliet Dolostone generally have had much more to fear from rapacious real estate developers and the city's mania for playing wrecking-ball bingo than from their crumbling over the course of geologic time into giant heaps of spalling flakes.

Above Old St. Pat's sturdy stone plinth rises its other vintage building material, Cream City Brick. This famous pale-yellow product of Milwaukee's Quaternary lacustrine and glacial clays often takes on a darker and dirtier hue when exposed to the urban atmosphere, as it does here in patches. But once again, this can be an ornamental asset. There are some lovely cream-brick churches to be seen in Wisconsin's lakeside cities that at time of writing are so majestically venerable in their sootiness that it would be a cardinal sin to clean them thoroughly. And I like Old St. Pat's just as it currently is, too, with its piebald brick and its two spires as unabashedly asymmetrical as the claws of a fiddler crab.

In this book, I restrict the name "Cream City" to brick made specifically in Milwaukee or its immediate vicinity—as the masonry here apparently was. However, cream-colored brick was also a very common nineteenth-century staple of other communities in southeastern Wisconsin and northeastern Illinois. Racine, for example, was a major production center, and even clay pits in smaller towns, from Port Washington and Kenosha to Dundee, Fort Sheridan, and Lake Forest, turned out pale-toned equivalents virtually indistinguishable from Milwaukee's. This was due mostly to the fact that tills and lakebed clays in that area have an usually high calcium and magnesium content that masks the effect of the iron also present. When fired, the clays' quirk of geochemistry expresses itself in white, yellow, and pale-peach tones, whereas the clays found elsewhere in the area, lower in magnesium and calcium, take on the deeper salmon and red, iron-oxide tints most commonly associated with American brick.

9.4   Holden Block

1029 W. Madison Street
Completed in 1872
Architect: Stephen Vaughan Shipman
Geologic feature: Buena Vista Siltstone

In and around the year 1880 there were two well-documented places in Chicago where one could see the Buena Vista Siltstone in architectural use. One was here on the Near West Side, at the Italianate commercial-loft building known as the Holden Block. The other was at the newly completed US Post Office and Customs House, located on the Loop block that now contains the Chicago Federal Center Plaza. But, as related in the "Foundations" section of part I, the Customs House met an untimely end due to its serious structural flaws. Its Buena Vista stonework (which most certainly did not cause the problems) was dismantled, shipped on 500 railroad flatcars to Milwaukee's South Side, and there was resurrected as the exterior of the imposing St. Josaphat Basilica.

Fortunately the more modestly scaled Holden Block, a preeminent example of early post–Great Fire construction that has its year of completion proudly emblazoned on the top of its façade, has survived. That, despite long spells of neglect, vacancy, and a generally precarious existence in the midst of much demolition nearby, during West Madison Street's "Skid Row" years. Recently it has been beautifully restored, and its surroundings have experienced a marked economic upturn.

The Buena Vista derives its name from one of the southern-Ohio communities where it has been quarried. As a siltstone, it resembles sandstone and can easily be mistaken for it, and especially for the buff-to-gray, medium-to-fine Berea Sandstone, another famous clastic rock type produced in Ohio and used in Chicago's post-fire rebuilding. The Buena Vista's grain size, usually in the coarse-silt category, can straddle the silt-sand nomenclatural divide. For that reason, anyone choosing to call it, like the Berea, a sandstone instead should hardly trigger our sniffy opprobrium. In my experience, though, its texture is less gritty and abrasive; the Berea was, after all, a favorite choice for grindstones. In any event, the Buena Vista is the younger of the two and dates to the lower Mississippian subperiod. In the nineteenth century, it was often given the honored epithet of freestone—a quarryman's term of admiration for any sedimentary rock type that possessed the highly desired and much appreciated quality of being easily worked in any direction without unwanted splitting. Most likely, the ornately carved stone on display here was quarried not in Buena Vista itself but in the Scioto County town of McDermott, just a few miles north of the Ohio River.

9.5 1508 W. Polk Street

Year of completion unknown
Architect: Unknown
Geologic feature: Artesian Dolostone

This small rowhouse, nestled in among its neighbors on a leafy residential street, was moved at some point from its original location on Ashland Avenue. Geologically, it's notable for its rock-faced ashlar, which is bitumen-spotted Artesian Dolostone that was quarried in the vicinity of the intersection of Western and Chicago Avenues, about 2 miles to the northwest of here. The earliest quarry in that locale apparently opened in 1849, with a handful of others joining it in the years following. Unfortunately, all were closed by the 1930s, used as garbage dumps, and eventually completely filled in. Now parks, factories, stores, and vacant lots stand where once there were great holes and headwalls exposing the city's Silurian bedrock.

The bitumen, a black, tarlike substance Illinois geologists have also called asphaltum, is a diagnostic characteristic of the dolostone extracted from the reefal rock of the Artesian quarries.

9.6   Church of the Epiphany

201 S. Ashland Avenue
Completed in 1885
Architectural firm: Burling & Whitehouse
Geologic feature: Jacobsville Sandstone

This site amply defends my proposition that, of any style on display in Chicago, the Richardsonian Romanesque provides the most immediate and sustained geological gratification, which is, of course, architecture's highest calling. This lovely house of worship, currently slated to be reborn as a performing-arts center, stands as a reminder of the days when designers reveled in the use of deep reds and browns and purples, before fashions flipped and light-toned stone, brick, and terra-cotta were selected with frightful regularity to enforce the dreary dogma of reason and cheerfulness. Later eras would confine richer, more characterful rock types to building bases and damp courses. Here and elsewhere the Richardsonians went full in and bet the lot on magisterial solidity and edifying gloom.

While in recent years restorers have discovered that some of the church's coping is an unknown red granite added to replace deteriorated original material, the ashlar that predominates here is the Jacobsville Sandstone, Michigan's version of Lake Superior Brownstone. This was the Midwest's most widely used answer to the famous Portland Sandstone quarried in Connecticut and Massachusetts that was so wildly popular in New York, Boston, and other Eastern cities. Today, the Church of the Epiphany is the best surviving documented example of our regional alternative employed in a grander public rather than private residential setting. It was apparently also Chicago's first use of this stone in rock-faced form—or so the *Daily Inter Ocean* reported in 1885.

Like the Portland, the Jacobsville was deposited in a major rift-valley structure, albeit in a much different time and place. Produced in the Upper Peninsula towns of Marquette, L'Anse, and Portage Entry, it could be shipped to Chicago easily by water and also by the rapidly expanding North Country rail lines. However useful and appealing it is when used architecturally, the Jacobsville

FIGURE 9.2. Purveyors of the Richardsonian Romanesque style in Chicago had a special predilection for "brownstone," and had quite a number of varieties from the Eastern US and the Midwest to choose from. The most popular of the lot, shown here in its variegated form at the Church of the Epiphany, was Upper Michigan's Jacobsville Sandstone.

is not without its persistent geologic mysteries. In common with northernmost Wisconsin's look-alike Chequamegon Sandstone, which may or may not be its temporal equivalent, the Jacobsville Sandstone lacks fossils and other forms of evidence that would give stratigraphers a better idea of its place on the time scale. For that reason, its age is, to use geo-jargon, "poorly constrained." Estimates have ranged from the late Mesoproterozoic to the Neoproterozoic and on into the early Cambrian, an uncertainty gap of over half a billion years. In any case, these sandstone layers were deposited after a great magma plume, like those now under Yellowstone and Hawaii, had risen from the Earth's mantle to create both a massive outpouring of basaltic lava and the Midcontinent Rift, a great trenchlike depression that stretched from what is now Oklahoma through Iowa and Minnesota to the Lake Superior region, and then back down to Alabama. Then sediments eroded from the nearby highlands were deposited in layer after layer over the hardened lava. Sitting at the top of this sedimentary sequence is the Jacobsville Sandstone.

9.7   First Baptist Congregational Church

60 N. Ashland Avenue
Original name: Union Park Congregational Church
Completed in 1871
Architect: Gurdon P. Randall
Geologic feature: Lemont-Joliet Dolostone

Like Old St. Peter's about a mile to the east, this church is a time capsule from Chicago's pre–Great Fire era. Architectural writers extol its unusually spacious amphitheater of a nave, but to the geologist its lofty Gothic stone exterior is the big story. For here is a first-class exposure of the city's favored building stone of that era: Lemont-Joliet Dolostone. Unlike the more locally quarried Artesian Dolostone on display at 1508 W. Polk, the yellow-weathering Silurian carbonate rock here is not spotted with black bitumen—though on my last visit I did note an anthropogenic equivalent in the form of some dribbled tar, perhaps from a recent roofing project.

For the most part, the stone that was produced in the Lower Des Plaines Valley towns of Lemont and Joliet was taken from the Sugar Run Formation rather than the reefal and interreefal, bitumen-spotted Racine Formation above it. Of the Sugar Run quarried at Lemont, Henry A. Bannister noted in the 1882 *Economical Geology of Illinois*, "The rock is a fine-grained, even-textured limestone, of an agreeable, light-drab color, when first taken from the quarry, and rubs well, though not capable of receiving a very fine polish. . . . By the exposure to the air it changes to a pale yellow, or buff color, which appears to be deepened by the smoky atmosphere of a city." Bannister also explained that "its accessibility to Chicago, and its general excellence as an ornamental stone, have made it almost the only material used . . . in that city, for facing outer walls, and for general outside decorative architecture. From its adaptability to these uses, it has fitly received the name of 'Athens marble,' by which it is known wherever it is used."

Nineteenth-century geologists generally called our Regional Silurian Dolostone "magnesian limestone," which in turn was often shortened, as it is above, to just "limestone," with no reference to its somewhat different chemical composition. This probably accounts for why most modern Chicagoans, engineers and architects certainly included, still call our native bedrock "limestone" instead of what it really is. Or perhaps "dolostone" is just one too many terms to retain. With regard to the marketing name cited by Bannister, the "marble" may have been an early example of advertising hype, but the "Athens" actually derives from Lemont's original name, and not from a certain Greek city which, as I recall from visits years ago, is also a place of some architectural distinction.

9.8   Metropolitan Missionary Baptist Church

2151 W. Washington Boulevard
Original name: Third Church of Christ, Scientist
Completed in 1901
Architect: Hugh M. G. Garden
Geologic features: Tiffany Enameled Brick, American Terra Cotta

When it was first built, this Greek Revival church was a showplace for Tiffany Enameled Brick, a burnt-clay product that had previously been mostly relegated to hallways, washrooms, and the interior light wells of such notable early sky-scrapers as the Loop's Marquette Building. Here, however, it finally proved its worthiness as a bona fide façade material. This new architectural application was heralded in such trade journals as *Brick* and the *Brickbuilder*. The latter reported that "Mr. Tiffany, the president of the company, has long contended that the use of enameled brick for exteriors is desirable, especially in many of the western cities, for many reasons, among them, and one of the principal, being that a building so built is better able to withstand the smoke and dirt nuisance and so preserve its original beauty. Another good reason is that rich effects may be obtained in the body of a building by the use of soft tints under a semi-glaze."

The Tiffany firm, based some 50 miles due south of Chicago in Momence, Illinois, used for its line of enameled products base material dug in the Hoosier hamlet of Clay City, which was mixed with a fire clay obtained from Montezuma, on the Indiana bank of the Wabash River. This special blend was first made into dry-pressed bricks and burned in a kiln to create the "biscuits" onto which the enamel compound was applied. These were then burned again in separate kilns at the unusually high temperature of 2,400 degrees Fahrenheit. This ensured that the coating actually fused with the underlying clay rather than just forming a hard but brittle overlayer, as with normal glazes. Enameled ceramic surfaces resist crazing, or the development of an unsightly network of fine cracks, much more effectively than simple glazing does. While the façade here does show a few chips in places, it has held up remarkably well, and still retains its resplendent whiteness.

Complementing the Tiffany Enameled Brick is another clay-derived material: beautiful cream-yellow trim, window frames, column capitals, inscription panels, and other ornamental details made by the American Terra Cotta Company, located in what is now Crystal Lake, Illinois. Most of its source clay conveniently came from its own property, which lay on a thick blanket of Pleistocene Lemont Formation glacial till, deposited at the base of the Wisconsin ice sheet approximately 20 ka ago.

While this house of worship certainly is first and foremost a gleaming master-piece of fired clay, it does feature some stone as well. To find it, simply examine the steps and two great column shafts at the entrance. Close inspection reveals it's an igneous rock composed of dark and light mineral grains. Described in the sources merely as "granite," it has, particularly on the columns, a distinctly banded and wavy, gneissic texture. Unfortunately its locale of origin was apparently not recorded; its identity remains a mystery.

9.9 Our Lady of Sorrows Basilica

3121 W. Jackson Boulevard
Completed in 1902
Architects: Henry Engelbert, John F. Pope, and William J. Brinkman
Geologic features: Salem Limestone, Carrara Marble, Chicago Common Brick, Celadon Terra Cotta Roof Tile

Chicago has three Roman Catholic churches that have been honored by papal decree with the designation of basilica; this was the first to be built. It once presented a more symmetrical profile on Jackson Boulevard, but the western of its two original steeples was lost by a lightning strike and the fire it ignited in 1984.

What remains is still geologically impressive. The façade is southern Indiana's Salem ("Bedford") Limestone of Mississippian age, both in dressed-faced and rock-faced forms. But the basilica's designers, obeying a standard cost-saving practice, restricted the bulk of this relatively expensive ornamental cladding to the north-facing façade. As a result the building presents an utterly different aspect on its eastern elevation. There, with the exception of the Salem trim, ornament, and base, the walls are reddish-brown Chicago Common Brick, and this humblest of building materials, taken from local Pleistocene glacial-till deposits, proves itself more than worthy in this high holy context. Above it, the roofing is now mostly asphalt shingle or something similar in appearance. But in a few places—most visibly, the top of the still-extant steeple and the far eastern end of the transept facing Albany Avenue—the original red terra-cotta tile manufactured by the Celadon Company seems to have survived. Celadon, originally based in Alfred, New York, had a number of plants in different regions, and at the time Our Lady's tiles were made, one of them was in Ottawa, Illinois, on the northern fringe of our state's coal fields, which yielded high-quality clays as well as fossil fuels. The Ottawa operation supplied much of Celadon's orders to the Chicago market, so it's reasonable to assume that the basilica's original roofing began as Pennsylvanian-subperiod shales and underclays deposited when the Midwest lay on the equator and was locked deep within the supercontinent Pangaea.

If Our Lady of Sorrows is open during your visit, take a few minutes to admire its magnificent interior and especially its splendidly grandiose altar, which is made from top to bottom of choice Carrara Marble. This most famous of architectural rock types, in widespread use since the time of Christ, first formed as Jurassic limestone on a carbonate bank of the once-great Tethys Ocean. Well over 100 Ma later it was metamorphosed into its current, superbly polishable form in the Oligocene, during the tectonic upheaval that produced Italy's Apuan Alps.

9.10   Garfield Park Bandstand

Music Court Circle in Garfield Park
Completed in 1896
Architect: Joseph L. Silsbee
Geologic features: Murphy Marble, Venetian Glass, Copper

It's a distinct shame that this lovely variation on East Indian and Arabic themes has been allowed to fall into a dilapidated state. On a recent visit, I even found small trees growing in its open interior. But most of the geologic materials the Bandshell was originally blessed with are still there, and in good shape.

The structure itself is built of the sugary, coarse-grained "Georgia White" variety of Murphy Marble, a limestone probably of Cambrian or Ordovician age that was metamorphosed later in the Paleozoic. In places where water drips down from above the stone hosts colonies of calciphilic algae, but generally it has held up well here. Laid within it are mosaic designs made of glass produced in La Serenissima, the lagoonal city of Venice, Italy, and specifically on its island of Murano. Above them the magnificent canopy retains its beautiful sheathing. An 1897 *Inland Architect* description of the Bandstand noted, "When the copper gets its natural oxidization it will last for a lifetime." Actually, the oxidizing is just the first step to the formation of the current copper-salts patina. Were you to watch a time-lapse movie of the structure from 1896 to present, you'd probably see a progression from the bright reddish metal to dark chocolate brown, and then a gradual transition to the bright green of today. And at an age of about 125 years, the copper here has already outlasted the journalist's prediction.

9.11   Lincoln the Railsplitter Statue

Garfield Park at W. Washington Boulevard and N. Central Park Avenue
Completed in 1911
Sculptor: Charles J. Mulligan
Geologic features: Berlin Rhyolite, Bronze

One of many bronze statues in Chicago endowed with a brown patina that is both attractive and protective, the Railsplitter stands proudly on a stone pedestal with rock-faced edges that conveys a sense of rustic ruggedness. Close examination reveals that the base is made of very hard extrusive igneous rock consisting of a purplish-black groundmass dotted with pink feldspar phenocrysts. This is the porphyritic Berlin Rhyolite, quarried in the Badger State town of that name (pronounced BER-lin, and not like the German capital). In the late 1800s a great quantity of this almost supernaturally tough stone was used as pavers for Chicago streets. Now, however, it's most frequently seen in cemeteries, since it was favored as an enduring and appropriately somber-toned selection for tombstones and grave monuments. But the chunk of stone you see here was deemed a project worthy of special mention in the June 22, 1911, issue of the *Berlin Weekly Journal*. There it was duly reported that this hefty mass of rock, taken from a locally quarried block originally measuring 5 x 5 x 5 feet and weighing 13 tons, had been finished and shipped to the big city, so that it and its statue would be ready for the official unveiling ceremony on the following Fourth of July.

While the Berlin Rhyolite excels in conveying a sense of permanence and solidity, its origin was overtly catastrophic. Late in the Paleoproterozoic era, about 1.75 Ga, the region that is now central and southern Wisconsin experienced a series of huge volcanic eruptions. Geologists have hypothesized that these explosive episodes were triggered by the thinning of the crust after an episode of continental collision and the uplift of the lofty Penokean Mountains. The eruptions included the release of immense amounts of superheated volcanic ash in the form of highly destructive pyroclastic flows, of the sort produced by Mount St. Helens in 1980 and Mount Vesuvius in 79 CE. While the Berlin Rhyolite was first thought to have formed from very viscous lava with the consistency of bread dough, it has more recently been reinterpreted as the product of congealed ash instead. So it might be better termed an ignimbrite or ash-flow tuff.

Another odd thing about the Berlin stone and other Wisconsin rhyolites of this era is their very dark color. Rocks of this type are most often light gray, pink, or brick red. Still, the mineral content of the Berlin stone is certainly what one would expect. It contains roughly equal proportions of quartz, which accounts for its hardness, and two feldspars, orthoclase and albite.

9.12 Garfield Park Fieldhouse

100 N. Central Park Avenue

Completed in 1928

Architectural firm: Michaelsen & Rognstad

Geologic features: American Terra Cotta, Flemish Bond Brick, Ludowici
Terra Cotta Roof Tile

There's gentle irony in the fact that Chicago's most dazzling example of the flamboyant Churrigueresque variant of the Spanish Baroque was designed by two
Norwegian American architects. The building's Hispano-Nordic style reaches
its crescendo at the main entrance, where the unsuspecting first-time visitor is
ambushed in a riot of flowers, seashells, busts, masks, lions' heads, volutes, twisting columns, and even a full-length figure of French explorer Robert Cavalier
de La Salle—with the whole lot being topped off with a central dome covered in
gold leaf, as well as green Ludowici Terra Cotta Tile on the shallow-pitched roofs

FIGURE 9.3. The Garfield Park Fieldhouse features intricate terra-cotta ornament in its most hypercaffeinated form. This manic assemblage of decorative
doodads was produced by the talented artisans of the American works, located
in McHenry County dairy country. Note also the building's beautiful Flemish Bond
brickwork.

of the end pavilions. This sublime mania for fiddly bits was rendered magnificently in over 285 tons of precisely aligned cladding panels manufactured by the American Terra Cotta Corporation. This firm, which was situated in the Illinois community of Terra Cotta (now part of Crystal Lake), was only able to compete with Chicago's own giant of the industry, Northwestern Terra Cotta, because of its lower product pricing. This in turn was the result of leaner labor costs in rural McHenry County, and of its on-site source of clay derived from Pleistocene glacial deposits. American was also obviously successful in attracting and retaining superb artisans who routinely turned out exceptional work in their remote, dairy country setting.

Just as impressive in its own way is the Fieldhouse's handsome brickwork. The cream-colored facing brick (its source was apparently not recorded) is set in Flemish Bond, a simple but harmonious and visually engaging pattern in which each course, or horizontal row, is composed of alternating stretchers (the long sides of bricks set horizontally) and headers (the short sides of the same). Note how the former of one course are centered atop the latter of the course just below it.

9.13   Humboldt Park Receptory and Stable

3015 W. Division Street
Current name: National Museum of Puerto Rican Arts and Culture
Completed in 1896
Architectural firm: Frommann & Jebsen
Geologic features: Ludowici Terra Cotta Roof Tile, Fieldstone, Salem
    Limestone

This whimsically eclectic structure, a combination of Queen Anne and German domestic styles, was intended to serve as the reception and stabling facility for park guests arriving in horse-drawn carriages. When it was built, it was an outstanding example of the ornamental potential of burnt clay as a roofing material. However, in later decades the tiles were replaced by infinitely less appealing asphalt shingles. Fortunately, the building's two phases of renovation in the 1990s included the reinstallation of magnificent red Ludowici Terra Cotta Roof Tile, in both traditional unglazed and glistening glazed forms.

Much of the exterior wall area is made of an unidentified red pressed brick; the old reliable, Salem ("Bedford") Limestone, is also present as the buttresses, sills, and other highlights. But the most eye-catching building material here forms the Receptory's distinctive base—a superb example of Fieldstone construction. The rounded to angular boulders include dark mafic and light-toned felsic igneous

FIGURE 9.4. A showplace of Ludowici Terra Cotta Roof Tile in both glazed and unglazed forms, the Humboldt Park Receptory also gives the urban geologist a fine opportunity to inspect glacial erratics, a gift of the Pleistocene ice age, used in the building's fieldstone base.

rocks from northern sources, and are most likely glacial erratics—separate chunks of rock transported to our region by the ice sheet of the final, Wisconsin glaciation of the Pleistocene epoch. Most of the largest boulders (and there are some very hefty ones) are set at the base, with the least bulky rocks placed at the top. This gives the building's bottom section an appearance reminiscent of the natural graded-bed pattern found in some clastic sedimentary strata, where the biggest and heaviest particles carried by a stream are deposited first, followed by the medium, and finally by the smallest and lightest.

9.14   Alexander von Humboldt Monument

Humboldt Park (N. Sacramento Avenue between North Avenue and
    Division Street)
Completed in 1892
Artist: Felix Gorling
Geologic features: Freeport Granite, Bronze

Independent explorer, geologist, science writer, botanist, zoologist, ethnologist, meteorologist, nonstop talker, feted celebrity and more, Alexander von Humboldt (1769–1859) was humankind's closest approach to a one-man civilization. While few who pass this site today are familiar with his life or legacy, the fact remains that his writings, adventures, and findings captivated generations of readers and inspired many of the greatest scientists who followed him. Humboldt, arguably history's greatest naturalist, was the very model of a self-starter intellectual on a lifelong quest for broad-based, multidisciplinary knowledge. Sadly, that's now largely a forgotten ideal subsumed by scientific hyperspecialization, rigidly defined academic castes, and mandatory institutional sanction.

I suspect that if Humboldt were still alive today, he'd be duly fascinated by his bronze statue here and its brown, copper-oxide patina that shields the metal inside it from further corrosion. It might even remind him of the protective, varnishlike coating he observed on granite boulders he encountered at the cataracts of the Orinoco River, during his great South American expedition. And, being a student of granites themselves, he'd probably also be pleased to discover that the statue's base is fashioned from a very rarely encountered example of this igneous rock type—the Freeport Granite. This light-bluish-gray stone has a remarkably fine texture and equally sized grains of microcline, orthoclase, quartz, oligoclase, biotite, and muscovite. Quarried in the seacoast town of Freeport, Maine, it began as a body of magma that rose higher into the crust from its place of origin. When this happened, though, is still uncertain. The most recent bedrock map of the Freeport Quadrangle assigns it, rather tentatively, to the long interval from the Devonian to the Permian.

9.15   Humboldt Park Boathouse

N. Sacramento Avenue between North Avenue and Division Street
Completed in 1907
Architects: Richard E. Schmidt, Garden & Martin
Geologic features: Stucco, Chicago Common Brick, Ludowici Terra
  Cotta Roof Tile

Just to the northeast of the Humboldt Monument stands this graceful, green-roofed essay in Prairie School horizontality. Its brick arch piers and lower exterior walls are made of brick that appears, from its range of colors and pitted texture, to be Chicago Common, made from our region's Pleistocene glacial till. The masonry above it is coated in handsomely understated buff stucco, the decorative outdoor version of plaster composed of Portland Cement, lime, sand, and water. And the roof itself, like its counterpart atop the Humboldt Park Receptory, is composed of terra-cotta tile produced by that leader of its industry, the Ludowici Company. Originally based in Chicago, the firm is now located in New Lexington, Ohio.

SOUTH LOOP, MUSEUM CAMPUS, PRAIRIE AVENUE, DOUGLAS, AND BRONZEVILLE

10.1 Congress Plaza Hotel

504 S. Michigan Avenue

Former names: Auditorium Annex, Congress Hotel

Completed in 1893; southern additions, 1902 and 1907

Architects: Clinton J. Warren (1893), Holabird & Roche (1902 and 1907)

Foundation: Shallow and spread (1893), wooden piles and caissons (1902 and 1907)

Geologic features: Mount Airy Granodiorite, St. Cloud Area Granite, Graniteville Granite

The main ground-floor cladding on either side of the hotel's Michigan Avenue entrance is the light-gray Mount Airy Granodiorite, of Mississippian age and quarried in North Carolina. This is the stone that was also chosen to replace the Aon Center's failing Carrara Marble. It's contrasted in the window bases by the darker-shaded St. Cloud Area Granite from Minnesota, here represented by its "St. Cloud Gray" variety. Dating to the Paleoproterozoic, it certainly seems to be the same sort of crystalline rock, yet it's five times as old as the Mount Airy. And the pavers of the entrance itself are made of a third granite, the "Missouri Red" variety of the Graniteville, which formed in the Mesoproterozoic, in between the other two on the geologic time line.

MAP 10.1. Chicago's South Side.

One can see in this trio of granitoids an example of remarkable continuity through vertiginous gulfs of time. From the early days of the Proterozoic eon to the latter half of the Paleozoic era, through all the changing dispositions of the continents, the shrinking and swelling of ocean basins, the comings and goings of simple and complex lifeforms coping with dramatic swings in global climate, the Earth's lithosphere has operated as a remarkably consistent machine, churning out vast quantities of new igneous rock in the same basic way for a billion years and well beyond that. Magma, hotter and less dense than the surrounding rock, has risen through the crust time and time again, like globs of wax in a lava lamp. And some miles beneath whatever was then the surface it has slowly cooled and hardened into the mosaics of feldspars, quartz, and other silicate minerals you see here.

10.2   Blackstone Hotel

636 S. Michigan Avenue
Completed in 1910
Architectural firm: Marshall & Fox
Foundation: Rock caissons
Geologic features: Salisbury Granite, Northwestern Terra Cotta,
   St. Louis Brick

This historic Beaux Arts hotel has, in terms of its exterior materials, a tripartite aspect: a squat ground-floor segment of stone, a somewhat more expansive midriff of enameled white terra-cotta, and an upper portion that is mostly composed of handsome red brick.

The ground-story stone is the "Balfour Pink" variety of Salisbury Granite, quarried near the North Carolina town of that name. Unlike the Mount Airy Granodiorite visible just up the block at the Congress Plaza Hotel, the Salisbury dates to the lower Devonian period, and is derived from magma that apparently rose all the way from the mantle and was emplaced in the upper crust, probably as part of the Acadian mountain-building event. It's approximately half pink orthoclase, with lesser amounts of white plagioclase feldspar, quartz, and biotite mica.

But the geology doesn't stop on the first floor. The clay that forms the baked bisque of the sparkling Northwestern Terra Cotta almost certainly had its source in the Pennsylvanian-subperiod shales and underclays of Illinois Basin coal country, either in this state or Indiana. It first formed mud layers and soil profiles in a climate decidedly more tropical than our own, at a time when the American Midwest, part of the immense Pangaean supercontinent, lay athwart the equator.

FIGURE 10.1.   The synergy of stone and clay. The South Loop's Blackstone Hotel, as lovely as it is famous, is an effective blend of geologically derived ornamental materials: a base of pink Salisbury Granite, a midriff of white Northwestern Terra Cotta, and a much more extensive upper portion of handsome red St. Louis Hydraulic Pressed Brick.

And the Blackstone's expanse of premium-quality facing brick is yet another story writ in the pages of geologic time and natural processes. It was produced by the St. Louis Hydraulic Press Brick Company, a giant of the industry capable of cranking out millions of highest-quality bricks each year in a wide range of colors, including the all-time favorite cherry red you see here. Its facing-brick clay was mined from abundant and geologically recent Mississippi River floodplain deposits, certainly no older than the Pleistocene epoch, that provided top-notch material. Interestingly, the greater St. Louis metropolitan area also has its own Pennsylvanian bedrock. It in turn provided the city's vast array of brickyards with the perfect clay used in pottery and heat-resistant refractory bricks destined for the linings of furnaces and kilns.

10.3   Second Studebaker Building

623 S. Wabash Avenue
Subsequent names: Brunswick Building, now the Wabash Campus
    Building of Columbia College
Completed in 1896

Architect: Solon S. Beman
Foundation: Shallow and spread
Geologic features: Andes Black Gabbro, Chicago Common Brick

At some point in the late 1950s the lower portion of this building was redone in a style that seems to have been maliciously calculated to be as jarring and disrespectful to the original design as humanly possible. The worst offending element, a set of aluminum fins, was later removed, to the special regret of this lifelong connoisseur of the ridiculous. What remains, however, is a geologically if not architecturally splendid exposure of the Andes Black Gabbro, a rather mysterious if once fairly widely used "black granite" selection. It has here been polished to such a state of reflectivity that on a sunny day one could comb one's hair in it. It seems the Andes Black was quarried in Brazil, in some locale in that huge country that is now reportedly a national park. Unfortunately, the details of the stone's geologic age and origin have not been widely disseminated. Still, being a gabbro, it is a thing of beauty in itself, with large and easily scrutinized crystals of pyroxene and plagioclase feldspar. It also supposedly contains traces of pyrite and magnetite.

On the southern side of the building, where the cladding ends, there's a far less showy but similarly significant surface to explore. The attraction here is not stone but instead Chicago Common Brick, caught residing in its preferred habitat, on a side or back wall where fancier facing brick was deemed unnecessary. When last I checked, it was quite sooty, but in places its yellow, salmon, and red tones still poked cheerfully through the grime.

The Chicago Common was always intended to be softer, more porous, and less precisely shaped than the expensive pressed bricks employed on building fronts. As architect George Beaumont wrote in 1886, it did its humbler job just fine by being "rough, crooked and fairly durable." Beaumont also noted that clay pits on the north side of town produced bricks both better shaped and richer in lime than those on the south. Various other sources confirm that in most cases, Chicago Common Brick was made from the region's abundant supply of Pleistocene glacial till, which is much more pebbly than the clays used for pressed brick. The latter derived instead from well-sorted Holocene stream deposits and the Pennsylvanian-age sedimentary shales and underclays transported by rail from Illinois Basin coal fields.

10.4   General John Logan Memorial

Grant Park (S. Michigan Avenue and E. 9th Street)
Completed in 1897

Architect: Stanford White

Artists: Augustus Saint-Gaudens (Logan figure), Alexander Phimister Proctor (horse)

Geologic features: Cape Ann Granite, Concrete, Bronze

A protégé of Ulysses S. Grant and William Tecumseh Sherman, Illinois native John Logan was a rare example of a political-appointee general who was an effective commander in the American Civil War. Later he successfully resumed his legislative career as a congressional representative, US senator, vice-presidential nominee, and commander of the very influential Grand Army of the Republic organization. Chicago's Logan Square is also named in his honor.

Like a general seeking the most advantageous high ground on a battlefield, architect Sanford White positioned the heroically posed bronze equestrian statue on a specially constructed hill, where visitors entering from Michigan Avenue see Logan and his high-stepping horse silhouetted against the sky like mythic figures. However, if one approaches from the north, this icon of the old martial virtues now has a futuristic, sci-fi backdrop of hivelike high-rise condos. The vision of master sculptors is juxtaposed jarringly against an alien backdrop embodying values of a very different sort, if any.

A multilayered geology lesson awaits at the hill's summit. Proceeding from top to bottom, there's the bronze ensemble of man and horse, a product of the Henri Bonnard Bronze Works in New York City. It has been given the brown copper-oxide patina also seen on the Equestrian Indian statues a few blocks up. Then there's the stone base, made from the "Bay View Green" variety of Massachusetts' Cape Ann Granite. As its trade name suggests, it was marketed as a green granite, but that tint is a subtle one that is here best recognized on the stone's polished section.

The concrete walkway around the base is of great significance, too, at least if one is a rockhound. It features a coarse, nonskid aggregate of surf-rounded pebbles. In their composition they closely match the medley of rock types found on many of this region's Lake Michigan beaches. Their lot includes Archean gray granite from the Canadian Shield, black basalt from the vast Proterozoic lava flows of Lake Superior's Midcontinent Rift zone, bone-white dolostone, and chert the color of butterscotch. The last two are relatively youthful emissaries from the Silurian strata, less than half a billion years old, that form the bedrock here and along much of Wisconsin's eastern flank. All this treasure is trodden on ten thousand times a year and otherwise left unheeded. But these everyday materials are intensely interesting arrangements of matter. They have an ancient earthly legacy. Only as an afterthought were they transported by a Pleistocene glacier and then caressed by its meltwaters. If one has been blessed with the curiosity to look into it, this legacy humbles, terrifies, electrifies, inspires. In the alert mind and activated soul a simple sidewalk becomes a surface sparkling with diamonds.

10.5   Field Museum

1400 S. Lake Shore Drive
Completed in 1920
Architectural firm: Graham, Anderson, Probst & White
Foundation: Wooden piles
Geologic feature: Murphy Marble

From a little distance, and particularly on an overcast day, the Field Museum seems to be the hundred-and-umpteenth example of a major Chicago building clad in Salem Limestone. But when you walk right up to it, you'll see it clearly isn't. In cleaner spots the stone is white, not gray, and it doesn't have the fine-grained feel of the Indiana rock. Instead, it's composed of coarse and interlocking calcite crystals devoid of the Salem's fossil content. This isn't limestone. It's a true marble—and specifically the justly renowned Murphy Marble. Initially, the stone here was all of the "Georgia White" variety; later it was replaced in spots by the very similar "Cherokee" brand. Both of these are paler equivalents of the veiny pink "Etowah" type used not all that far away at Buckingham Fountain.

For some reason the feel of this nonfoliated metamorphic stone really appeals to me. I've been known to run my fingers caressingly over it, in a way only another geologist could fully forgive. Recently, when I was showing a little too much lithic love interest in one of the sections of the museum's southern side, I was actually apprehended by the Chicago Police Department, in the person of a bomb-sniffing springer spaniel named K9 Buster. After ascertaining that the perp was not explosive, Buster called in backup in the form of his human patrol partner, Officer J. Stephen, who was attached to the other end of his very long leash. Once I explained to both woman and dog that I was researching a book on the city's architecture and was thus intrinsically harmless, I was cleared of all charges and had a good chat with my two new friends about the local geology.

An impressive amount of the Peach State's upper crust has gone into the Field's exterior. The museum's website states that 350,000 cubic feet of its marble were used in the original construction. While over the past century it has shown the effects of weathering, it has not suffered the overtly grisly fate experienced by the Italian Carrara Marble used in the Aon Center. However, deteriorating terra-cotta ornament and ashlar block shifting due to corroding iron anchors necessitated a two-year restoration here in the 1980s that involved the installation of about 350 tons of new stone that was quarried just two miles from the original Pickens County, Georgia, source. A product of the Murphy Marble Belt about 40 miles north of downtown Atlanta, this rock type has been in use in Chicago since at least the 1880s. First laid down as limestone in the Cambrian period or thereabouts, it was metamorphosed into its current form later in the Paleozoic period, during successive spates of mountain building and continental collision.

10.6   Shedd Aquarium

1200 S. Lake Shore Drive
Completed in 1929
Architectural firm: Graham, Anderson, Probst & White
Foundation: Wooden piles
Geologic feature: Murphy Marble

What was said of the Field Museum can be repeated here. The exterior's handsome white stone is "Georgia White"—the Murphy Marble.

10.7   Adler Planetarium

1300 S. Lake Shore Drive
Completed in 1930
Architect: Ernest Grunsfeld
Geologic features: Morton Gneiss, Copper, Bronze

What once was a set of shoals off the Lake Michigan shoreline became, with the construction of the Adler Planetarium, Chicago's bridgehead to the rest of the universe. The exterior of this twelve-sided structure (each facet representing a sign of the zodiac) is clad in Morton Gneiss, the most appropriate rock type the architect could have chosen, given the high cost of meteorites. I say this because, at the ripe old age of 3.52 Ga, this Paleoarchean migmatite is both the most ancient stone in architectural use in America and the one that formed soonest after the 4.54 Ga birth of the Earth and the rest of the solar system. In fact, it's even a quarter the age of the entire cosmos.

Nowhere else in the city is this striking and magnificently complicated rock selection on such extensive display. See if you can hunt down the triangular clast of black amphibolite, about a foot long, that has a wavy line of light-toned granite transecting it. Its tight meanders and loops are *ptygmatic folds*, which are characteristic of highly metamorphosed rock where a fairly stiff and viscous material (here the granite) was dramatically contorted within a much more yielding host substance (the amphibolite). One of the best ways to conceptualize ptygmatic folds is to look at them as waves frozen in time, where the amplitude (the height from trough to crest) is much greater than the wavelength (the length from one trough to the next). Given the Morton's hardness now, it's difficult to imagine the hellish conditions that would have made these two ingredients flow, bend, or give way at all. But then, at three-quarters of the age of

FIGURE 10.2.   Lovely ptygmatic folding in the Adler Planetarium's Morton Gneiss cladding. Here a vein of younger granite wends its switchbacked way through a chunk of amphibolite. Dating to the far-distant Paleoarchean era, the Morton is the oldest stone type used in Chicago's architecture—and perhaps in all of architecture.

our planet, this rock has endured more traumas than the fleeting human mind can possibly imagine.

The Adler is also a showplace of copper extracted from our planet's crust. In the main doorway and the exterior's ornamental bas-relief panels it's present in alloy form, combined with tin to produce bronze. These panels were designed by sculptor Alfonso Iannelli and cast by the American Bronze Company. But the copper is also present in its elemental form, as the sheathing of plates that adorns the outside of the central dome. This malleable, highly conductive, and easily worked metal, which can be handily hammered or cut into desired shapes, has been in human use for at least 10 ka. Nowadays it's most frequently extracted from two sulfide-mineral ores, chalcocite and chalcopyrite, and it makes a very attractive and enduring, if also rather costly, roofing material. Proponents of "green" architecture are quick to point out its energy efficiency: it reflects rather than absorbs the ultraviolet part of solar radiation that strikes it. And even in a building dedicated to taking in as much of the rest of the universe as possible, that's a good thing.

10.8   Balbo Monument

Burnham Park, east of S. Museum Campus Drive, south of
   E. McFetridge Drive, west of Lakefront Trail
Erected in 1934
Architectural firm for the base: Capraro & Komar
Geologic features: Unidentified ancient marble, unidentified ancient
   breccia, Tivoli Travertine, differential soil settling

In Chicago architecture, it's easy to find the neoclassical; finding the non-neoclassical is a different matter. But here it is. This Corinthian column is a fascinating if lonely vestige of Imperial Roman architecture that now stands on an inland lakeshore of a continent unknown to its makers.

Regrettably, it has become the object of much vituperation, though the reason it's controversial is obvious enough. It bears the misfortune of having been Benito Mussolini's gift to this city at the time of its Century of Progress Exposition—a gift also intended to commemorate the 1933 Italy-to-Chicago flight by a squadron of seaplanes commanded by Italo Balbo. To make matters worse, the inscription on its base includes boastful references to Il Duce's special brand of *fascismo*. And so various Chicagoans at various times have denounced this artifact as an overt threat to democracy and called for its removal, despite its having had no impact on local political systems whatsoever. Were Chicago not a town widely respected for its utter lack of humbug one might suspect that some no-pain, all-gain grandstanding was going on.

While I have no more fondness for fascism than anyone else familiar with the tremendous grief and evil it has spawned, I do suffer from a different perspective—one the predominantly political mind must, I suppose, always resent. When I look at this little monument, a survivor of so many centuries, the pomposity and silliness of its latter-day inscription fade into the mists of archaeological and geologic time. It's possible to look beyond the words, at the column that once stood amid the hustle and bustle of the ancient Roman seaport of Ostia. Now it offers its services as an educative emissary to a much younger culture that still has much to learn. Let's recork our bottles of vitriol and examine this artifact in detail.

Its capital is made of white marble—possibly the Roman *marmor lunense* version of the Carrara. Or it could be the Pentelic or Parian varieties from Greece. In contrast, the shaft is composed of some sort of Mediterranean Basin breccia. The whole thing reportedly dates from the years 117–138 CE. That's interesting, because this is the exact span of the reign of Hadrian, that most architecturally activated of emperors. In his day, tastes in ornamental stone definitely ran toward the exotic and showy, and this breccia certainly qualifies as that. Though its badly weathered state makes it difficult to tell, some marble clasts and darker fragments (serpentinite?) are still visible within it. So I've wondered if it might be an early example of Greece's Larissa Ophicalcite that can also be seen, in much more recently quarried form, at 333 N. Michigan and the Elks National Memorial. However, the Romans also used breccias from North Africa and Asia Minor that might fit here, too.

The twentieth-century base is fashioned of the pitted Tivoli Travertine, the Quaternary spring-deposited limestone quarried east of Rome. Note also how the monument is gradually becoming a local approximation of the Leaning Tower of Pisa. This no doubt is due to differential settling of the soil or fill material beneath it. The angle of tilt is hardly alarming, but here on a small scale you can nevertheless see what the architects and foundation engineers of Chicago's larger buildings have always had to fret about. In Chicago as in Italy, one can never assume the *terra* will ever be *firma*.

10.9 Second Presbyterian Church

1936 S. Michigan Avenue
Completed in 1874, bell tower in 1884
Architects: James Renwick Jr. (1874), John Addison (1884)
Geologic feature: Artesian Dolostone

Northeastern Illinois currently boasts three architecturally notable churches built of the unusual blotched stone produced by the Artesian quarries of Chicago's West Side. They're Oak Park's Pilgrim Congregational, Lake Forest's First Presbyterian, and this lovely English Gothic edifice at the corner of Michigan Avenue and Cullerton Street.

The history of Chicago's Second Presbyterian extends back to the formation of its congregation in 1842. By 1847 the decision had been made to build the city's first stone church, at the intersection of Washington and Wabash, now a busy Loop locale occupied by such landmarks as the Marshall Field's store complex and the Pittsfield Building. Prominent New York architect James Renwick was engaged to design this house of worship in the same Gothic idiom he later used on a more grandiose scale for Manhattan's St. Patrick's Cathedral. This earlier incarnation of the Second Presbyterian was soon nicknamed "The Spotted Church" and even "the Church of the Holy Zebra"—references to the blobs of black bitumen, a naturally occurring asphalt compound, found on its Artesian Dolostone exterior. While this kind of feature might seem unsightly to the modern eye, it was at that time considered, as the architectural historian Daniel Bluestone has noted, a positive ornamental effect imparting an air of hallowed ancientness to the building—a view any geologist past or present could certainly agree with. Though later destroyed in the Great Fire of 1871, the Spotted Church still lives on in one sense, because much of its distinctively marked stone was salvaged and can now be seen 30 miles to the north, where it graces the exterior of the Lake Forest house of worship cited above.

Fortunately for the urban geologist, the current Second Presbyterian, another Renwick design, also features Artesian Dolostone, which the 1885 *Marquis' Hand-Book of Chicago* describes somewhat more poetically as "rock-faced prairie stone." Here too the exterior is dotted with bitumen; and once again the local press praised its venerable look.

When first exposed on a freshly broken surface, this viscous hydrocarbon goo often gives off a noticeable petroleum odor, but it later loses its distinctive bouquet and hardens to a crustlike consistency. Here care must be taken not to confuse it with the black soot staining that is most evident on the trim of the once-buff sandstone. (That sandstone, incidentally, is most likely the Berea, from Ohio. It's heresy to say so, but I think the soot looks good here and provides additional character and contrast.) The real bitumen is found only on the dolostone, and is distributed randomly in little clusters that look like splatters flicked from a carelessly wielded tar brush.

The Artesian quarry district that produced the ashlar for both this church and the others mentioned above was located in the vicinity of the intersection of Western and Chicago Avenues. Its name comes from the fact that there was a

FIGURE 10.3.   Stonework of the Second Presbyterian Church exterior. The predominant and paler rock type here is the bitumen-spotted Artesian Dolostone, quarried on Chicago's West Side. However, the smooth-faced course is most likely composed of either Berea Sandstone or Buena Vista Siltstone, two Ohio selections much in demand after the Great Fire and often used specifically for ornamental building trim. Both of these relatively porous clastic sedimentary rocks eagerly collect anthropogenic soot—a much more recent additive than the Artesian's blobs of ancient asphaltum.

well situated near the crossroads that was indeed artesian: the groundwater that came from it had the highly desirable property of being confined belowground under positive hydrostatic pressure. When tapped it willingly rose to the surface on its own, without the need for expensive or laborious pumping. This well was used to create a pond for ice making, an important industry in the years before electric refrigerators and freezers were introduced.

In the Artesian district several separate producers of dimension stone, lime, and crushed rock flourished for a while. Among them, at different times, were the Taylor, Giles, Rice, Chicago Stone and Mining, and Artesian Lime and Stone companies. If you visit this part of town nowadays you'll be hard pressed to find evidence that any of them ever existed; by the first half of the twentieth century all had closed and were then filled in with trash and other debris. Now their lots are mostly devoted to parks and commercial buildings. When they were active, however, these quarries busily extracted rock from the Racine Formation, the uppermost portion of the region's Silurian strata. The Racine is geologically and paleontologically renowned for its extensive reefal and interreefal deposits. These have yielded up amazing insights into the complex marine ecosystems that throve here, on a carbonate bank of the Kankakee Arch about 425 Ma ago. The reef rock in particular tends to be more resistant to erosion, coarser-textured, and vuggy (perforated with voids). It was into these holes in the rock that much of the bitumen, the last organic remains of countless ancient lifeforms from a later geologic period, seeped and was subsequently preserved.

10.10   John J. Glessner House

1800 S. Prairie Avenue
Completed in 1887
Architect: Henry Hobson Richardson
Geologic features: Milford Granite, Lemont-Joliet Dolostone, Valders
   Dolostone, Decorative Mortar, Akron Terra Cotta Roofing Tile,
   Chicago Common Brick

While buildings of the Richardsonian Romanesque style are easy to find in Chicago, those designed by Richardson himself are not. As a matter of fact, with the demise of the Marshall Field Wholesale Store and the American Express Company Building, both demolished in 1930, the Glessner House became the solitary surviving example. But it wonderfully demonstrates Richardson's predilection for rock-faced ashlar exteriors and his happy blending of geologically derived materials. And if ever a family residence has deserved the grand honorific of Urban Geologic Wonder, this grand mass of magnificent stone and burnt clay certainly does.

From the street, the eye-catching exterior is composed of Milford Granite, quarried in the Milford-Braggville locale of southeastern Massachusetts. Neoproterozoic in age, it formed as part of a batholith in the far-wandering microcontinent of Avalonia long before it collided with what is now eastern North America. Chicago is a showcase of many types, textures, and hues of granite, but the Milford, with its coarse-grained salmon matrix sprinkled with dark flecks, is unquestionably one of the most attractive and immediately recognizable. At this site, it's completely accessible for close examination. A little scrutiny with a hand magnifier reveals a constellation of interlocking crystals: pink orthoclase and microcline, pale-blue quartz, white to yellowish-green albite, and black biotite mica. And notice how remarkably effective the brick-red decorative mortar between the blocks is in highlighting them.

Interestingly, architect Richardson's own choice for the stone above the house's base was the pink "Etowah" version of Georgia's Murphy Marble. After his untimely death, but before construction had proceeded to the point it was actually used, the Glessner family, who'd seen the "Etowah" used on another Chicago residence—very possibly the Weiss House, described in this book—decided they preferred the Milford for the entire exterior.

This site is also the place to pay one's respects to the most frequently overlooked application of stone in building construction: as the pavers we walk on but rarely notice. The buff-colored sidewalk slabs flanking the house are not original elements; they date just to 1978, but are of special note because they're made of Valders Dolostone. It's a far-northern representative of the Regional Silurian Dolostone complex, and one that, unlike its Illinois equivalents, is still quarried for architectural use. Its small town of origin, Valders, lies due west of Manitowoc, Wisconsin, and about 70 miles north of Milwaukee. (As I write this, I turn to a nicely sawed and shaped sample of this rock that the Valders Stone and Marble Company kindly sent to me. This chemically precipitated sedimentary rock I hold is dense, compact, smooth but not slick, and uniformly unassuming in its off-whiteness. Furthermore, when I place a drop of dilute hydrochloric acid on it, it resolutely refuses to react. It doesn't produce even a single bubble of carbon dioxide. So it truly is a dolostone rather than a limestone—an excellent and enduring choice for the job it's been given here.)

Fortunately, the Valders can be compared to one of our own region's Silurian dolostone equivalents, the Lemont-Joliet. The latter is visible in the trim and lintels of the house's inner courtyard. In this more secluded setting note how satisfactorily Chicago Common Brick takes the place of the Milford Granite, which was reserved for public display. While this was not the place for fancy, expensive facing brick, the Chicago Common here was carefully selected for quality and a fairly tight range of pink tints.

Gloriously topping everything described so far is another Richardson specialty: a roof of terra-cotta tile. Originally it was all of the Akron variety, manufactured in that Ohio city by J. C. Ewart & Company. Its source clay came from Akron's own, glaciated Summit County and possibly other parts of the state where Pleistocene till was a common source material for brick- and tilemakers. Now the Glessner roof is a composite of those original tiles and identical replacements salvaged from the Former Chicago Historical Society building. Regardless of their origin, they're the same color as the decorative mortar used to bind the stone below them, and they complement both the Milford and the Chicago Common superbly. Just imagine, in contrast to this chromatic harmony, the less happy marriage that would have resulted from a roof covered with gray slate or glazed green tiles instead. Just as it is, the Glessner House has everything in its proper place.

10.11   William W. Kimball House

1801 S. Prairie Avenue
Current name: US Soccer Federation Headquarters
Completed in 1892
Architect: Solon S. Beman
Geologic feature: Salem Limestone

Standing directly across the street from the imposingly granitic Glessner House, the Kimball manse should be given the award for Best Residential Use of Salem Limestone. Untold megatons of this rock type, also known as "Indiana Limestone" and "Bedford Stone," can be found in the Loop and everywhere else in the city, soaring into the clouds on Art Deco skyscrapers, forming entire hulking exteriors of Gothic churches and the becolumned Beaux Arts façades of train stations, museums, and public libraries. The geologist taking all of this in is forced to wonder how the upper crust of southern Indiana can still possibly exist, so much has been taken from it. But then one sees the Salem here, in a much more intimate setting, so beautifully and intricately carved.

It's true that some of the sculpted detail lower down on the façade has suffered over the years from weathering and is less sharply delineated than it was originally, but much else on the house still looks well defined. What's of equal significance is that in places nature's artistry is also apparent, in the form of the stone's slanted crossbedding patterns. These indicate the direction and vigorous sediment-shifting action of water currents in the shallow carbonate shoals of the Mississippian sea covering much of the Midwest 340 Ma ago.

10.12   Elbridge G. Keith House

1900 S. Prairie Avenue
Completed in 1870, Mansard third story added in 1880
Architects: Jonathan W. Roberts (1870), unknown (1880)
Geologic feature: Lemont-Joliet Dolostone

The oldest surviving residence in this historic district, the Keith House (or rather, its lower two-thirds) antedates the Great 1871 Fire and stands as an example of the early popularity and widespread availability of Lemont-Joliet Dolostone, the beautiful though later much maligned carbonate sedimentary rock quarried in the Lower Des Plaines Valley to the southwest of the city. Its center of production was particularly well situated because it was adjacent to the Illinois and Michigan Canal and competing rail lines that took advantage of the same natural transportation corridor, which a few thousand years before had been a great drainageway for Glacial Lake Chicago. There on the floor and bluffs of this broad, flood-scoured stream channel the bedrock lies at or very close to the surface; there is no thick blanket of overburden, no deep layers of outwash or till, to laboriously excavate first.

After the Great Fire, architects were disappointed by our local Silurian dolostone's poor performance in the conflagration—it often cracked and crumbled under great thermal stress. They also complained that when exposed to the elements the rock tended to spall and scale. As a result, the building trades largely turned to other selections. Ultimately the Salem Limestone became their almost monolithic favorite, even though its depressingly uniform bluish-gray aspect is all too reminiscent of buck-naked Portland Cement. Our local rock, so beautiful in its varied textures and buttery-yellow weathering tones, was held almost beneath contempt. Only later was it fully realized that many varieties of sedimentary rock, and not just the local stuff, perform poorly when not quarried correctly, seasoned sufficiently, or set in the most advantageous orientation by experienced masons who understand the special quirks and needs of each stone type.

Geologist and building-stone expert George Perkins Merrill, writing in 1891, suggested that most of the weathering woes of Chicagoland's dolostones were due to their being hammered with heavy tools that "stunned" them and thus created microfractures that encouraged later disintegration. It also became better understood throughout the building industry that very few stone types, whether igneous, sedimentary, or metamorphic, are really all that resistant to fire. For that attribute, one simply had to turn to brick and terra-cotta, which, after all, are born in the fire and blistering heat of the kiln.

While the sides of the Keith House are walls of exposed brick, the main façade is clad in Lemont-Joliet Dolostone, and it shows it off to magnificent effect. It seems to have held up very well here in this secluded spot, though restoration efforts in recent decades may have added to its current pristine appearance. In any case, this residence and a large collection of other older buildings in the city demonstrate that the Lemont-Joliet and other selections of our Regional Silurian Dolostone complex have generally stood the test of time much better than their late nineteenth-century detractors led us to believe they would.

10.13   Calvin T. Wheeler House

2020 S. Calumet Avenue
Alternative name: Joseph A. Kohn House
Completed in 1870
Architect: O. L. Wheelock
Geologic features: Temiscouata Slate, Copper

Like the Keith mansion on nearby Prairie Avenue, the Wheeler House dates from just before that benchmark event in Windy City history, the Great Fire of 1871. Fortunately, this neighborhood stands considerably to the south of where it and the less well known but still very serious conflagration of 1874 raged. The first- and second-floor bay windows on the southern side of the façade, which were added later, are framed in copper, that most beautiful of metals, and the one with the longest record of human use. When I last checked here, it sported different colors in different places: the characteristic lurid reddish glint of fresh metal, as well as the pinks and greens and browns of oxide, carbonate, and sulfate weathering products.

The source of the Wheeler House's handsome red facing brick has not been documented, but if it's original it may well have been made in Philadelphia, at that time the nation's brickmaking supercenter, or in Baltimore. The roof above it, however, is much better provenanced, having been redone in recent years in the "North Country Unfading Black" variety of Temiscouata Slate. While it's often positioned, as here, where the urban geologist can't inspect it at close range, slate is still worth appreciating. A foliated metamorphic rock whose parent is shale, which in turn is derived from clay-sized sediments, slate is hard, enduring, but quite obligingly fissile. In other words, it splits reliably, with little coaxing necessary, into thin sections. This makes it a top choice for natural roofing material.

The Temiscouata Slate is lower Devonian in age, so each and every bit of it secured to this mansard roof is guaranteed by its makers, the forces of marine deposition and continental collision, to be about 400 Ma old. It was taken from the Glendyne Quarry, in Saint-Marc-du-Lac-Long, Quebec. Apparently this rock began its eventful life as beds of turbidite (marine avalanche) sediments in the foreland basin that lay in front of the developing Acadian Mountains. Close at hand, it's actually a dark gray, but when installed on a roof, it does look black, especially on an overcast or rainy day. Like so many of the other building materials on display in a city, this stone has a surprisingly dramatic life story. It began in murky waters off a now-vanished mountain range, eventually formed the bedrock of the glacier-scoured woodlands of the Quebecois outback, and then ended up here, atop an elegant Second Empire mansion in America's great midland metropolis.

10.14   3300 Block, S. Calumet Avenue

Geologic features: Regional Silurian Dolostone, Oneota Dolostone, Rock-Faced Brick, Roman Brick, unidentified red sandstone, Napoleon Sandstone, Salem Limestone

Many South Side residential blocks that retain their older houses are geological treasure chests waiting to be unlocked and appreciated. The virtue of this one small section of the Douglas neighborhood, intended as a sort of sampler of what's to be found over a larger area, is that it features an usually diverse selection of identifiable stone types in one compact locale. The five sites described below, all on the western side of the street and all Richardsonian Romanesque or a variant thereof, are by no means the only sparkling gems you'll uncover on this block in the course of a quarter-hour stroll.

George H. Edbrooke Houses, 3314 and 3316 S. Calumet (1884, George H. Edbrooke)

Both the larger main house and its adjunct are clad in Chicagoland's native rock type, Regional Silurian Dolostone. Its nonreefal, thin-bedded texture suggests that it was quarried from the Sugar Run Formation of the Lemont-Joliet district, but other possibilities can't be ruled out without good documentation, which is lacking here.

Clarence A. Knight House, 3322 S. Calumet (1891, Flanders & Zimmerman)

This essay in shades of orange is a masonry masterpiece that boasts a lower façade of slightly pinkish, sunset-tinted Oneota Dolostone, marketed as "Kasota Stone," and an upper portion of deeper-toned, Rock-Faced, Roman Brick. The Ordovician-age Oneota, which usually can be reliably identified by its color, is in Chicago most famously used in River North's Cable House. But its role as a striking and eminently sculptable ashlar is evident here as well.

John B. Cohrs House, 3356 S. Calumet (1890, architect unknown)

Here's the mystery house of the set. Its rock-faced ashlar, while immediately suggestive of the widely used Lake Superior varieties produced in Wisconsin and Michigan, could have been quarried instead in Connecticut, Massachusetts, Maryland, Colorado, southern Illinois, or Iowa. All these states contributed red

FIGURE 10.4.   The city's South Side is a vastly underappreciated geological treasure chest. Take, for instance, the Douglas neighborhood's 3300 block of S. Calumet Avenue. Its splendid collection of stone residences includes, from left to right, the Seaton House (Salem Limestone), the Southard House (Napoleon Sandstone), and the Cohrs House (unprovenanced brownstone).

or maroon sandstones rich in iron oxides to the manic building boom of post–Great Fire Chicago. But wherever this stone came from, it makes a lovely ruddy contrast to its two neighbors just to the south, by striking the deepest note in this trio's progressive chromatic shift from dark to light.

Albert R. Southard House, 3358 S. Calumet (1890, architect unknown)

If the Knight House's Oneota Dolostone is recognizable just by its color, this site's Napoleon Sandstone, a slightly eerie greenish yellow, is even more vividly so. Dating to the Mississippian subperiod, the Napoleon hails from Jackson County in Michigan's Lower Peninsula. In its role as a building stone it was most frequently listed as "Michigan Buffstone" or "Michigan Green Buff," though at least one modern source uses the geologically incorrect term "Greenstone." That's a tag best reserved for metamorphosed basalt even if Windy City architectural historians happily fling it at any rocklike substance of the appropriate color.

At time of writing this most immediately eye-catching residence of the row is undergoing a major renovation. The geologists and rockhounds among us can only hope and pray that the builders and new owners retain its rare and remarkably mustard-hued stone exterior. The Napoleon Sandstone can also be found on the Lincoln Park Presbyterian Church, but here its visual impact is even greater.

Chauncey E. Seaton House, 3360 S. Calumet
(1890, architect unknown)

It may be faced in our most relentlessly common rock type, Indiana's Salem Limestone, but what a magnificently uncommon job of intricate detail has been sculpted into it. Better known as "Bedford Stone," this Mississippian-age biocalcarenite here amply demonstrates why, toward the end of the nineteenth century and well into the twentieth, the Salem outcompeted every other stone type in the construction market. Here and on some other houses on this block it provides a chaste and cheerful counterpoint to the coarser textures and more lurid stone choices of neighboring homes. On the Seaton House, it's displayed in both smooth-sawn and rock-faced forms.

10.15   George M. Gartside House

663 E. Groveland Park Avenue
Completed in 1885
Architect: Willett L. Carroll
Geologic feature: Euclid Siltstone

The Gartside House is part of the Groveland Park residential complex that is tucked in between Cottage Grove Avenue and the South Shore tracks in the Douglas neighborhood. Several of the other homes here are worthy of interest because they're faced in such laudable materials as Lemont-Joliet Dolostone and a dusky sandstone that is either Lake Superior Brownstone or something very similar in appearance.

Still, it's the Gartside residence, shy and retiring in its secluded spot at the end of the row, that's the real star of the lot. For its front is adorned with a rock type one would now be hard pressed to see used as cladding elsewhere in this city, though I live in hope that someone will prove me wrong on that point. This is northeastern Ohio's Devonian-period Euclid Siltstone. There was a day when this rock type actually was a common sight in Chicago, in the sense it was considered a leading choice for sidewalk pavers and flagging.

In the building boom after the 1871 Great Fire it was most often called "Euclid bluestone" or just "Euclid stone."

Contemporary issues of the *Inland Architect* also bear witness to its use here as ornamental building trim. But to see this much of it on display as the main element of a façade is most unusual.

The Euclid Siltstone does have one Achilles' heel noted by nineteenth-century geologists and architects: it sometimes contains deposits of pyrite, otherwise known as fool's gold. This iron-sulfide mineral breaks down rapidly in the presence of humid air—something, as all Chicagoans know, we have in plentiful supply—and leaves unsightly rust stains. This chemical reaction can also affect the integrity of the rock and cause it to blister and peel. Here at the Gartside House I've seen no evidence of this going on, so I'd guess the stone was selected with care. But what is interesting is its complex coloration, an interplay of yellowish buff and palest grayish-blue. The former tint may well be a weathering product; the latter no doubt accounts for the rock's common name. (In the nineteenth-century stone trade, "blue" was a term bandied about with considerable artistic license, whether one was referring to the Euclid or to a grayer variety of Indiana's Salem Limestone.) While it may not be able to compete chromatically with lapis lazuli, the Euclid Siltstone is lovely to look at in its own right, and it's a notable addition to our city's roster of rarer rocks.

10.16   Stephen A. Douglas Tomb

636 E. 35th Street
Completed in 1881
Architect: Leonard W. Volk
Geologic features: Vinalhaven Granite, Hallowell Granite, Lemont-
  Joliet Dolostone, Shelburne Marble, Bronze

Stephen Douglas, the "Little Giant," was unquestionably one of the most prominent of nineteenth-century Prairie State politicians. While he died just one year after losing the 1860 presidential election to Abraham Lincoln, the completion of this grandiose tomb and 96-foot-tall memorial was delayed for two decades due to funding problems. Originally, the lower section that contains the sarcophagus was composed entirely of Lemont-Joliet Dolostone taken from the Silurian bedrock southwest of the city. Now all that remains of that initial stone choice is the one great paver slab that forms the door sill leading into the burial chamber. What has replaced it everywhere else on the exterior, with the exception of the five bronze statues, is a pair of nicely contrasting granites quarried in Maine.

The first of these is the "Fox Island" brand of Vinalhaven Granite, which is used to very good effect in a variety of finishes on the monument's lower level, including the stand-alone plinths for the bronze allegorical figures that silently watch over their surroundings. This igneous intrusive rock is a product of the rugged, wave-swept seacoast southwest of Acadia National Park. It formed as one of a swarm of Silurian plutons that were emplaced in what had been the microcontinent of Ganderia. In the late Ordovician to early Silurian that landmass had collided with Laurentia, the Paleozoic version of North America. A particularly attractive and enduring stone selection, the Vinalhaven comprises large crystals of buff orthoclase, dark-gray quartz, white oligoclase, and black biotite.

The upper portion of the monument, including the 46-foot column and its own plinth, offers a marked difference in color and texture. It's the Hallowell, the granite most likely to be mistaken for Salem Limestone from a distance. When one is more than a few steps away from it, it seems to be a uniform pale gray because of its small grain size. In the Chicago of the late 1800s the Hallowell was the most widely advertised and effectively marketed stone of its type. And it was especially favored for the monuments of the rich and powerful, as can be seen at Graceland and Rosehill Cemeteries. Quarried in the inland town of the same name, the Hallowell is Devonian in age, and came into being during the Acadian Orogeny that was triggered by Laurentia's collision with yet another microcontinent, Avalonia.

Even when the door grating is closed and locked, one can peer in and get a good look at both the sarcophagus that holds Douglas's remains and the sculpted bust of him that rests atop it. These are fashioned from Shelburne Marble, quarried from the lower-Ordovician bedrock on the western side of the Green Mountain State. Douglas was a Vermont native, so this choice is most appropriate, and here it's at least somewhat sheltered from the blast and acid content of the elements. It has thereby avoided the partial dissolution and sugared texturing endured by other Shelburne Marble graveyard monuments exposed to the full brunt of Chicago's ravaging weather.

10.17   D. Harry Hammer House

3656 S. Martin Luther King Drive
Completed in 1885
Architect: William W. Clay
Geologic features: Unidentified brown sandstone, unidentified red
    face brick, Copper

No roster of Chicago's most notable houses would be complete without this paragon of Bronzeville's stately homes. A paragon indeed, but of all the sites described in this book it's probably the most endangered, and certainly the most heartbreaking to see. At time of writing it has been allowed to fall into a deplorable state. This eye-catching Queen Anne beauty, once the busy home of famous community activists Lutrelle and Jorja Palmer, is now empty and forlorn, with boarded windows and crumbling brownstone. It needs a savior, and it needs one now.

The somber maroon foundation and decorative details may be one of the Lake Superior sandstones quarried in Wisconsin and Michigan's Upper Peninsula, or possibly even the Portland from New England's Connecticut River Valley. Regardless, it is spalling badly, particularly around the main entrance, where it may have been face-mounted. This was a widespread nineteenth-century practice of mounting ashlar sections perpendicular to their natural bedding plane that emphasized the consistency of a stone's initial appearance over its long-term survivability.

The red facing brick, still so striking despite the neglect, is similarly undocumented. At least superficially in better shape than the brownstone, it provides a handsome contrast to darker rock and the bright blue weathered copper turrets and trim. Were it brought back to its former glory, the Hammer House would make an outstanding museum or community center worthy of this historic and geologically fascinating neighborhood.

10.18   Chicago Bee Building

3647 S. State Street
Completed in 1931
Architect: Z. Erol Smith
Geologic feature: Northwestern Terra Cotta

Originally headquarters for a newspaper serving Chicago's African American community, this Art Deco edifice is now one of Bronzeville's public library branches. Its exterior is best seen in its entirety from across South State when the street trees fronting it aren't in leaf. Then the full glory of its Northwestern Terra Cotta façade, with its harmonious palette of pale green, medium green, black, and gold, can be appreciated in a single glance. And at much closer range you'll marvel at the workmanship and artistry of the stylized botanical motifs that adorn the design.

The Northwestern works, located along the North Branch of the Chicago River, obtained its high-quality clays from the coal fields of the Illinois Basin.

FIGURE 10.5.   One of Bronzeville's Art Deco gems. The Chicago Bee Building, once the headquarters of a prominent newspaper published for the city's African American community, is now a branch of the Chicago Public Library. It's elegantly clad in Northwestern Terra Cotta in a striking range of colors.

Accordingly, the base material for the cladding panels here was originally either a nearshore-marine mud deposit or the paleosol (ancient soil) in which the luxuriant equatorial vegetation of what is now the American Midwest grew. Miners call these latter beds underclays; they still often contain the root traces of the giant trees that dominated the Pennsylvanian coal-swamp ecosystems of 300 Ma ago.

MCKINLEY PARK, BACK OF THE YARDS, KENWOOD, WASHINGTON PARK, HYDE PARK, WOODLAWN, AND ENGLEWOOD

11.1 William McKinley Memorial

Northwest corner of McKinley Park
Completed in 1905
Architectural firm: Pond & Pond
Sculptor: Charles Mulligan
Geologic features: Concord Granite, Bronze

Erected to honor the memory of the twenty-fifth US president, who was assassinated in 1901, this memorial's bronze statue is the work of the same sculptor who designed Garfield Park's Lincoln the Railsplitter. Here, unlike there, the figure has been allowed to develop a classic blue-green patina indicative of copper carbonate or copper sulfate, with the latter probably more likely, given the urban and industrial pollution this locale has seen since the beginning of the twentieth century.

The statue's pedestal and the surrounding exedra are made of Devonian-period granite excavated from Rattlesnake Hill in Concord, New Hampshire. The Concord Granite owes its existence to a body of magma that rose into the upper crust and solidified there during the formation of the Acadian Mountains. It was also the time of the world's first forests and the earliest vertebrate land animals.

It was from that far-distant version of the Earth this light-gray, faintly blue-tinted intrusive igneous rock has come to this corner of the Windy City. And from the much more recent time when it was first quarried, it's been highly prized by American architects, especially for external use in monuments, public libraries, and commercial buildings of the grander sort. When last I checked here, the stone was in need of a good if careful scrubbing. and was not in the best shape for detailed scrutiny of its constituent micas, quartz, and feldspars. But even when a bit sooty, the McKinley Memorial is a very good place to appreciate the Concord Granite's durability and architectural attributes.

11.2   Union Stockyards Gate

850 W. Exchange Avenue
Alternative name: The Old Stone Gate
Completed in 1879
Architectural firm: Probably Burnham & Root (see the description below)
Geologic features: Lemont-Joliet Dolostone, Copper

The design of this sole survivor of the once extensive Union Stockyards has long been attributed to the architectural partnership of Burnham & Root. While direct evidence of their involvement here is lacking, most of the other structures that once stood in this famous complex were definitely their work.

I can remember the unequivocal, death-affirming bouquet the stockyards produced. Given the help of a malevolent southerly breeze, the stench could make it up to the Loop and other neighborhoods as well. But I did not actually visit the stockyards proper, or what remained of them, until 2005, which was 34 years after this once-colossal operation shut down forever. But when I finally did make it here, I was heartened to find another surviving example of Lemont-Joliet Dolostone architecture. And I confess I was morbidly fascinated by how far the destructive process of weathering had actually gone in places on the gate (and continues to go, more than a decade and a half later). The structure was obviously in no danger of crumbling and falling to the ground in front of me, but in various spots it was shedding like the proverbial sheepdog. Little piles of recently spalled-off stone crust lay about on the sidewalk; and some of the ornamental details looked as if they'd been half nibbled away by some voracious species of mutant stone-eating moth.

The gate's dolostone was quarried specifically in the town of Lemont, a place of good people, rich history, lovely big-valley views, and beautiful, usually more enduring stone. The rock strata there are mostly part of the Sugar Run Formation. Ever since the Silurian, when these beds formed as layers of calcareous mud in the subtropical shallows of the Kankakee Arch, perched between two deeper basins, they've been locked tight in the Earth's crust. There they've been

FIGURE 11.1.   The sole surviving architectural icon of a bygone era, the Union Stockyards Gate remains one of the Windy City's great examples of its original preferred rock type, the Lemont-Joliet Dolostone.

protected from the wild swings of temperature found on the surface and from the ravages of acid-laced rainwater. But then, in this last century and a half, the stone you see here has reentered the fast lane of the Earth's grand Rock Cycle. Taking the geologist's long view, an act certain to infuriate every good architectural preservationist, one could say the dolostone is just doing what it was put on this planet to do: to change into something else when the conditions are right. Other Lemont stone—perhaps well seasoned or set better, or placed in a less polluted location—can be found in much fitter shape on various other Chicago landmarks. Why not repurpose the Union Stockyard Gate as the new Chicago Museum of Mutability and Natural Processes? I'd graciously accept the position of Curator of Crumbling and Disintegration Science, and spend my work days attending longwinded staff meetings and designing innovative hands-on learning activities for visiting school groups.

For those not quite so enchanted by the necessity of things falling apart, the gate does display one much more pleasant and less messy manifestation of weathering. It's to be found on the steeply pitched roof, which since 1971 has been sheathed in copper intentionally oxidized into a dark brownish-gray patina. By preventing still more corrosion, this sort of change actually maintains the chemical status quo.

11.3   Joseph H. Howard House

4801 S. Kimbark Avenue
Completed in 1891
Architectural firm: Patton & Fisher
Geologic feature: New York Red Slate

Of all the houses in the Kenwood neighborhood that feature slate as a building material, this exemplar of the Queen Anne style was the one that caught my eye the most and prompted me to consult with Joseph Jenkins, author of *The Slate Roof Bible*. Joe agreed with my initial assessment that, while the roof pitches are now covered with asphalt shingles or a similar material, the vertical surfaces are still clad in tiles of what is almost certainly New York Red Slate. Most are set in a distinctive fish-scale pattern.

The New York Red is the most striking of several color varieties of slate quarried in Washington County, New York, and adjoining Vermont communities for roofing tile and quite a few other uses. All began as deep-water ocean mud deposits off the southeastern coast of Laurentia (ancestral North America) in the Ordovician period. These deposits were thrust up by an advancing arc of volcanic islands onto what was then our continent's margin, as part of the formation of the ancient Taconic Mountains. In the process of being bulldozed ashore, they were folded and metamorphosed into their present form. The New York Red Slate's striking color signals the presence of the iron-oxide mineral hematite.

11.4   Blackstone Public Library Branch

4904 S. Lake Park Avenue
Completed in 1904, children's annex in 1939
Architect: Solon S. Beman
Geologic features: Concord Granite, Salem Limestone, Carrara Marble, Holston Limestone, Larissa Ophicalcite

This august if modestly scaled building is the crown jewel of the Kenwood community. It's also a geologist's joy to explore, inside and out. The cool, bluish-gray exterior consists of Concord Granite which, like the stone of the McKinley Memorial, came from Rattlesnake Hill, on the northwest side of New Hampshire's capital. This Devonian igneous intrusive rock has never had a better showplace. It seems to have been expressly made by the forces of nature for the classical formalism of the Beaux Arts style. Here it's fashioned into pilasters, Ionic columns, and various other Greco-Roman doodads, from urns to acroteria. My favorites are the projecting palmettes near the entrance which, like the rest of the ornament here, bravely defy any functional significance whatsoever. This little shrine to the goddess of learning takes itself very seriously. As it should.

If you pause a moment to look closely at the Concord Granite here, you'll enter the crystalline world of interlocking silicate minerals. The atoms that now compose these chemical compounds had once belonged to sediments and ocean floor that were thrust in a subduction zone deep into the Earth's interior. Finally they were melted by friction into a body of magma. Less dense than the crust around it, this great mass of molten rock rose back toward the surface but cooled before reaching it. And so were created the feldspars, micas, and quartz that now form the library's outer walls. See if you can spot the tiny plates of sparkling muscovite that are one of this fine-grained granite's chief identification traits. You may also notice some spalling of the stone on the eastern side. Even such tough rock types as the Concord are bound to suffer some deterioration when exposed to an environment saturated with acidic rain, dramatic swings in temperature, and deicer salts.

On venturing over to the children's annex at the southern end of the library complex you'll notice that its exterior is clad in a rock type of a warmer, buff tone. This is Indiana's widely used Salem Limestone.

Happily, the Blackstone Branch is one of those relatively few Chicago buildings that also has some well-documented interior stonework. You'll first see it in its amazing, muralled mini-rotunda. Its walls, like those in other rooms of the library, are clad in white and gray-veined Carrara Marble. In use since ancient Roman times, it's been quarried in massive quantities from the Apuan Alps of northern Italy. Formed from Jurassic limestone that was metamorphosed in the Oligocene, the Carrara remains to this day the most prestigious of all statuary-grade marbles.

At the base of the rotunda lies one of the city's most beautiful walking surfaces, a harmonious combination of mosaic work and other fancy stone selections. The two rock types here that have been cited in the sources are the "Champion Pink Tennessee Marble" variety of the Ordovician Holston Limestone (at center) and Greece's Larissa Ophicalcite (sea green with large clasts of lighter marble; often called "Verde Antico Marble"). I'm tempted to guess what the lovely but unrecorded yellow stone between them is, but there are too many options offered by Old World quarries to make my guess worthy of print.

11.5   Francis M. Drexel Fountain

S. Drexel Boulevard and E. 51st Street
Completed in 1881, enlarged in 1888

Sculptor: Henry Manger
Geologic features: North Jay Granite, Bronze

The bronze figure perched high atop an ornate pedestal commemorates Philadelphia banker Francis Martin Drexel, who donated a portion of his South Side real estate holdings to the city of Chicago. While this ensemble's alloy of copper and tin (here coated with the characteristic green, copper-carbonate and copper-sulfate patina) is always of interest to the urban naturalist intrigued by the chemistry of weathering, the primary geologic lure is the wide stone basin beneath. This is composed of Maine's North Jay Granite, often touted as one of the whitest of its kind in the stone trade. This is the best place in town to see it in good shape, with all of its minerals present and accounted for. At its other primary exposure, on the 1892 section of the Loop's Marshall Field's Store, it is in a considerably more degraded condition.

Devonian in age, the North Jay formed from a large mass of magma emplaced in the upper crust during the Acadian Orogeny. According to the New England granite expert T. Nelson Dale, it owes its strikingly pale shade to its combination of translucent, untinted quartz and the fact that its two alkali feldspars, microcline and orthoclase, are both bone-white. Its mineral complement also includes the plagioclase feldspar oligoclase and two micas, black biotite, and silvery muscovite.

11.6 Hyde Park Union Church

5600 S. Woodlawn Avenue
Completed in 1906
Original name: Hyde Park Baptist Church
Architect: James Gamble Rogers
Geologic features: Jacobsville Sandstone, Salem Limestone

There are not many stone types used in Chicago that immediately offer up their identity to the urban geologist—even to one zipping by in a car without stopping. But this is one of them. The arresting brick-red exterior of this church shouts out the presence of the distinctive "Portage Red" variety of Michigan's Jacobsville Sandstone, which in turn is garnished with buff-colored Salem ("Bedford," "Indiana") Limestone trim.

Because the Salem is amply discussed elsewhere, our focus here is on the "Portage Red." This version of the Jacobsville owes its trade name to its locale of quarrying, the small Keweenaw Peninsula community of Portage Entry. Like the other Lake Superior Brownstone selections once used for architectural stone in

Chicago, Milwaukee, Duluth, and other Midwestern cities, the age of this rock type has not been definitively determined. Unfortunately, it lacks the fossils and other features geochronologists rely upon for more precise dating. Over the last century, estimates have therefore ranged all the way from the Mesoproterozoic to the Cambrian, with the early Neoproterozoic being an especially plausible hypothesis. In any event, the Jacobsville Sandstone owes its origin to the deposition of sandy sediments eroded from surrounding highlands after the late Mesoproterozoic's great Midcontinent Rift event. That infernal episode of crustal thinning, rising hot-spot magma, and intense volcanic activity almost resulted in the breakup of our continent.

One common if not ubiquitous feature of Lake Superior Brownstone, in both its Michigan Jacobsville form and Wisconsin's more-or-less equivalent Chequamegon Sandstone, is its variegation, caused by the presence of white streaks and pale reduction zones, circular to oblong areas in the rock. Probably the streaks and some of the wider spots, too, formed when groundwater solutions selectively leached out or reacted with the stone's preexisting hematite (iron-oxide) content while still in the ground. But at least some of the reduction zones instead seem to be places where an enclosed pebble or organic compounds kept the sandstone from oxidizing as it did so abundantly elsewhere. You can see good examples of variegated stone by the church's main entrance.

11.7 Frederick C. Robie House

5757 S. Woodlawn Avenue
Completed in 1909
Architect: Frank Lloyd Wright
Geologic features: Roman Brick, St. Louis Brick, Ludowici Terra Cotta
    Roof Tile, Decorative Mortar, Salem Limestone

This world-famous statement of the Prairie Style is, among many other things, Chicago's best exposure of Roman Brick. It's also a compelling demonstration of how three different geologically derived building materials, when cleverly combined, can work synergistically to produce a stunning visual effect.

For this project architect Wright traveled to St. Louis to choose the exact variety of Roman Brick he wanted. Some modern sources call it "Pennsylvania Iron Spot," though apparently this was just a trade name employed decades later for brick of very similar appearance. At any rate, the Robie brick features black flecking in a matrix of earthen reddish-brown, and this coloration adds to the impression that here the architect's palette harmonizes rather than competes with the surrounding landscape. To produce the flecking, brickmakers had to fire the

FIGURE 11.2. A stroll about Hyde Park's world-famous Robie House quickly reveals what a profoundly geological mind its designer, Frank Lloyd Wright, truly had. The architectural equivalent of a sequence of sedimentary strata exposed on a river bank, the building's thicker bands of Salem Limestone trim alternate with slender courses of Roman Brick, cleverly mortared to emphasize their horizontality.

clay at a higher temperature than normal, to coax the iron to congregate on the surface. As attractive as the result was, this brick type like all others has not been invulnerable to weathering and chemical interactions with the mortar. Recently some of it has been replaced due to its deterioration by exact replicas made by an Ohio firm.

At this juncture it should be pointed out that what modern architects call Roman Brick isn't exactly the same size as the type used by the ancient Romans themselves. For that reason one modern historian argues that this term should be avoided in describing this site in particular. But I here follow the practice of the majority of his colleagues. As currently defined, Roman Brick is longer, deeper, and shallower than the size most commonly used today. At the Robie House specifically, it's 1–5/8 x 4 x 11–5/8 inches, compared to the usual 2–1/4 x 3–5/8 x 8. All this might seem like a dry recitation of fractions, but it does specifically quantify why the bricks impart a perceptible sense of greater horizontality. Wright further accentuated this effect by specifying that the masons use decorative mortar of two colors: red in the vertical joints between the bricks, and white carefully recessed or "raked out" on the horizontal ones, for greater, shadowed relief. So from even a short distance away the courses appear to be continuous slim strata.

The long, low-pitched roof sections are covered in the same material found on many nearby University of Chicago buildings—Ludowici Terra Cotta Tile in its classic red color, which provides a harmonious complement to the brick and stone below. And for trim, Wright chose the neutral-gray Salem Limestone, also known as "Bedford" and "Indiana Limestone." Far and away the most commonly used building stone in Chicago, this Mississippian-age biocalcarenite is the perfect base and capping for the much ruddier exterior walls. Many other brick buildings in the city also have Salem trim, but nowhere else is this standard cliché of rock types used with more precision, impact, or cogency.

11.8   William Rainey Harper Memorial Library

1116 E. 59th Street
Completed in 1912
Architectural firm: Shepley, Rutan & Coolidge
Geologic features: Salem Limestone, Ludowici Terra Cotta Roof Tile

Geology, like the English language itself, is an almost endless collection of synonyms and alternative names. Three specialists looking at the same rock will shamelessly call it three different things, and with little or no prompting will pull additional labels out of their terminological toolkits. For example, the building material featured here, on the exterior of this academic quadrangle complex, can be identified as the Salem, "Bedford," or "Indiana" Limestone; or as grainstone, biocalcarenite, or freestone. This plurality of maddening monikers simply reflects the fact that this one rock type can be seen from many different perspectives. And the same thing is true of certain places discussed in this book. In this locale, for instance, what many would recognize rather prosaically as the University of Chicago I describe in a more geologic vein as the Great Hyde Park Carbonate Deposit.

The twin-towered Harper Library makes an ideal starting point for an exploration of this major geologic landform, which can be seen from space, or at least from a passing helicopter. Its Collegiate Gothic walls, clad in the standard ivy outfit, are just one small section of the vast assemblage of Salem Limestone and red Ludowici Terra Cotta Roof Tile that stretches, in other predictable Gothic arrangements, across the expanse of the campus. In 1894 the *Architectural Record* noted, "As this style of . . . architecture easily takes on the air of age by the help of a few vines and weather stains, the effect will certainly be most restful and suggestive of university conditions."

The cement-gray, fossiliferous rock of this megastructure formed in the Mississippian part of the Carboniferous period, when its quarrying locale of southern Indiana resembled what is now the Bahama Banks. But an immense amount of those ancient carbonate shoals has come to rest here and in the other older buildings of the campus, where it constitutes either a monument to the unity of an artistic vision or to dreadful lithic conformity, depending on one's viewpoint. Regardless, it's an impressive sight, especially when thousands of young backpacked bipeds scurry through the gaps in the encrusted calcite, like schools of fish negotiating the intricacies of a reef.

11.9   Rockefeller Memorial Chapel

5850 S. Woodlawn Avenue
Original name: University Chapel
Completed in 1928
Architect: Bertram Grosvenor Goodhue
Foundation: Rock caissons
Geologic feature: Salem Limestone

This 32,000-ton Gothic house of worship and campus spiritual center is the most impressive outcrop of the Great Hyde Park Carbonate Deposit. Here the Salem Limestone is load-bearing and sits on a stolid foundation of 53 caissons that reach the Silurian dolostone bedrock at about 80 feet down. With an ultimate compressive strength of at least 4,000 pounds per square inch, this fossiliferous sedimentary rock is definitely strong enough to bear the weight of these lofty walls and buttresses. The Salem is also what quarrymen call a freestone—it has no natural inclination to split in any preferred direction. As a result it can be easily cut, shaped, and manipulated just as one wishes. This quality and the rock's relative softness make it ideal for detailed carving. The ornate ornamentation, the work of sculptors Lee Lawrie and Ulric Ellerhusen, certainly bears witness to that.

11.10   University of Chicago Bookstore

5750 S. Ellis Avenue
Completed in 1902
Architectural firm: Shepley, Rutan & Coolidge
Geologic feature: Vermont Unfading Green Slate

This paragon of campus bookstores and safe haven of the mind offers among so many other treasures a rare glimpse of an interesting metamorphic rock type. That rock is definitely present elsewhere in the city, but it's rarely documented and almost always overlooked. Here as elsewhere it must be admired from afar, and you wouldn't be ill-advised to bring along a pair of field glasses. To see it as best you can, stand in the plaza directly to the south and gaze up at the bookstore's roof pitches. And there it is: the Vermont Unfading Green Slate.

This particular selection is just one of a broad spectrum of colors, from green and gray to purple and red, that is produced in the slate quarries of towns on both sides of the Vermont-New York border. All these variants have been used over the decades for everything from roofing and flooring tile to flagstones, old-style classroom blackboards, and billiard tables. Their different hues reflect subtle chemical differences in the environments in which the slate varieties formed. In the case of the Vermont Unfading Green, there is an abundance of the chlorite mineral group, which comes into being when biotite and other mafic minerals are subjected to the increased temperatures and pressure of metamorphism.

Regardless of their exact color or mineralogical makeup, Vermont and New York slates shared a common origin as deep-water ocean muds that were laid down in Neoproterozoic or Cambrian time. Then, in the Ordovician period, these deposits were thrust miles inland, folded, and metamorphosed when an arc of volcanic islands plowed into the margin of Laurentia, North America's forerunner. Huge slices of rock were scraped up and pushed across what is now the state of Vermont, to create the Taconic Mountains. Geologists call such units allochthonous (originating elsewhere; alien to their current setting). Now in places in the western side of the Green Mountain State one can find older types of rock that are perched atop younger. That's a sure tipoff that something tectonically wild and wonderful has happened.

11.11 Midway Studios

929 and 935 E. 60th Street
Completed in 1906
Architectural firm: Pond & Pond
Geologic features: Chicago Common Brick, Temiscouata Slate

The place to start at this cluster of buildings is the Gray Center Lab, at 929 E. 60th, facing the ominously looming mass of the Logan Center. Here that humble hero of Windy City architecture, the Chicago Common Brick, is on ample display, though it shares its stage with another, more uniformly red brick. All in all, this patchwork wall offers the student a three-credit course in brickology.

The urban field naturalist quickly learns that the Chicago Common can be distinguished ecologically—by the environments it inhabits. For older structures, these are primarily those out of the public gaze, on side and rear-facing walls, or on front façades of pretensionless buildings made purely for function. But in a habitat populated by younger structures, the rules take a vertiginous flip. There Chicago Common is most likely to be observed in recycled form, in the most favored and visible spots, as hipsterish statements of Retro Chic.

And then there's identification by morphology. First, the range of colors: pale yellow to salmon, but also redder. Then the shape, which lacks the sharp edges and relentlessly regular form of considerably more expensive pressed brick. And the texture: soft, with gaps and pits caused by the dolostone grit and pebbles that were embedded in the local clay used in brickmaking. When fired, these chunks of carbonate rock turned to clots of lime, which often subsequently popped out. Here you can also see one common weathering process that afflicts many types of brick exposed to the elements, and not just Chicago Common. Patches of white mineral crust, known as *efflorescence*, are the result of salt compounds in water solution that migrate through a porous substance—and brick is that—to evaporate on the surface. These salts often come from the mortar, but they may also be present in the brick clay itself.

Just to the east, at the Taft House section of the complex at 935 E. 60th, you'll see an exterior of a handsome if unidentified red face brick, and above it, a roof dressed out in the "North Country Unfading Black" brand of the Temiscouata Slate. It's a foliated and fissile Devonian metamorphic rock of Quebecois origin that's rich in carbonaceous minerals.

11.12   Reva and David Logan Center for the Arts

915 E. 60th Street
Completed in 2012
Architectural firm: Tod Williams Billie Tsien
Geologic feature: Bonneterre Dolostone

Of all the buildings described in this book, this one most successfully breaks the constraining bonds of symmetry, grace, and local context. Proof that the tower portion of this building actually originated on this planet is surprisingly evident in the form of its light-gray and buff-mottled cladding. At first look it seems to be Roman Brick on steroids, but in fact it's a truly attractive stone rarely seen in

Chicago, the "EW Gold" variety of the Bonneterre Dolostone. That's a good and earthly name if ever there was one.

The Bonneterre is quarried near Ste. Genevieve in southeastern Missouri, on the northeastern flank of the ancient St. Francois Mountains. It formed in shallow marine conditions, as layers of limey mud that were subsequently dolomitized—that is to say, infused with greater magnesium content. In this same environment, tropical trade winds caressed the tops of nearby stromatolite reefs at low tide. This was the time of the worldwide rise in sea level geologists call the Sauk Sequence. Our continent, which then straddled the equator after a long sojourn deep in the Southern Hemisphere, was oriented ninety degrees clockwise from its present disposition. And due to the Earth's greater rotational speed, each day was a little less than twenty-one hours long. Consequently, there were 424 days each year.

The Bonneterre Dolostone's economic importance in much more recent times is not just due to its use as dimension stone. It's also the main host of Missouri's ample deposits of galena and sphalerite, the primary ores of lead and zinc respectively. These ores accumulated in the Bonneterre after migrating a great distance in solution as hydrothermal fluids through more porous sandstones below the dolostone. This all occurred much later, in the lower Permian period, during the collision of South America with southern North America during the formation of Pangaea.

11.13   Deco Arts Building

1525 E. 55th Street
Original name: Ritz 55th Garage
Completed in 1929
Architect: M. Louis Kroman
Geologic features: Northwestern Terra Cotta, Milbank Area Granite

I've been intrigued and delighted by this building ever since I did my first architectural geology tour of the Hyde Park neighborhood two decades ago. As its current name suggests, it's Art Deco, and one of the city's best demonstrations of this style's use of terra-cotta. Not surprisingly, the cladding panels here were fabricated by Chicago's Northwestern Terra Cotta Company, which used as its source material Pennsylvanian-age underclays (Pangaean coal-swamp soil horizons) and shallow marine shales extracted and shipped in from strip-mined areas of the Illinois Basin. What Northwestern's artisans did with this byproduct of ancient ecosystems was always remarkable: here the Americo-mythic symbology includes stoplights, steering wheels, and entire long-snouted roadsters bearing Thoroughly Modern Millies replete with windblown scarves. You won't see this blithe-spirited approach to ornament in the stiff-necked neoclassical confines of lower LaSalle Street.

Obviously the terra-cotta is the main story here, but in fact there is a bit of lithic interest as well. Along with an undocumented gray stone type serving in places as a base course on the eastern side, there are two upward-flaring pillars flanking the automotive entrance at the building's southeastern corner. These are made of the richly reddish-brown "Agate" variety of Milbank Area Granite, quarried in Ortonville, Minnesota, an igneous stone's throw from the South Dakota border. Simply put, this rock is very old and more than half the age of our planet. Radiometrically dated to the Neoarchean era, it now and again shows banded, gneissic texture that indicates it was at least partially metamorphosed at some point in its unimaginably long history.

11.14　Museum of Science and Industry

5700 S. Lake Shore Drive

Former names: Palace of Fine Arts, Fine Arts Building, Field Columbian Museum, Field Museum of Natural History

Original version erected in 1893, conversion to a permanent museum, 1929–1940

Architects: Charles B. Atwood of D. H. Burnham & Company (1893), Graham, Anderson, Probst & White (1929–1940)

Geologic features: Salem Limestone, Yule Marble, Coal Mine Exhibit

For all their architectural grandeur, most of the structures of Chicago's epochal Columbian Exposition were not built to last. Their main building material was *staff*, a composite of plaster of Paris, hemp fiber, cement, and horsehair that did its job sufficiently for the needs of a world's fair open to the public for just six months. But among their number stood the Palace of Fine Arts, the immense neoclassical edifice that would eventually become the Museum of Science and Industry. It was made of somewhat more enduring stuff. Because it was intended to hold priceless works of art from around the world its staff exterior was complemented with internal steel trusses, iron columns, and fireproof inner walls composed of some 13 million bricks. Its current exterior of Salem Limestone, otherwise known as "Indiana" or "Bedford" Limestone, was installed decades later, with it fully in place in 1930. The expanse of this Mississippian-subperiod carbonate rock on display here presents itself in its familiar understated and grayish-buff aspect.

Inside the museum, a lesser-known but more overtly beautiful stone can be examined on the piers of the central dome. This is Colorado's splendid Yule Marble, which like the Salem is Mississippian in age. Originally part of the bluish-gray Leadville Limestone, it took on its current crystalline texture, and its white and sometimes handsomely veined appearance, when it was cooked in the heat of a nearby magma intrusion—a classic example of what geologists call localized contact metamorphism. This true marble, still quarried today in the heart of the Rocky Mountains, matches Italy's Carrara every bit in appearance and premium quality.

Anyone who has visited the museum also knows of another first-rate geologic point of interest, the Coal Mine, which in 1933 was the first exhibit opened to the public. A transgenerational favorite of Chicago families and school groups, it recreates Old Ben No. 17, one of countless deep-shaft longwall mines sunk into that vast sink of bituminous coal, the Illinois Basin. Such sites provided the cheap, abundant, sooty, and sulfurous energy that powered the Midwestern version of the Industrial Revolution in a time when few were aware of its terrifying long-term environmental and climatic impacts. Having paid my first visit to the Coal Mine over 60 years ago, I can vouch for the appeal of its simulated adventure; many years later, having descended in a real cage hoist 600 feet into Old Ben No. 25 in West Frankfort, Illinois, I can testify to the skill and accuracy of the museum's preparators. Now more than ever this is a place one should see to contemplate our civilization's choices past and present.

11.15 Statue of *The Republic*

Jackson Park (E. Hayes Drive and S. Richards Drive)
Erected in 1918
Sculptor: Daniel Chester French
Architect: Henry Bacon
Geologic features: Stony Creek Gneiss, Gilt Bronze

Standing majestically as a one-goddess greeting committee in a busy triangular intersection, this 24-foot-tall gilt bronze statue is actually a one-third-sized copy of the gigantic original that graced the 1893 World's Columbian Exposition. Its pedestal and base are made of Stony Creek Gneiss in a nonreflective rubbed finish. The Stony Creek is a rock type with a very complicated history that apparently includes chapters from both the Neoproterozoic and the Permian. If you review its geologic descriptions over the past century, you'll see it suffers from a triple-split identity crisis as a granite, gneiss, and migmatite. The stone you see here was taken from the Norcross quarry in Branford, Connecticut, which also was the source of the Newberry Library's exterior.

11.16   Antioch Missionary Baptist Church

415 W. Englewood Avenue
Original name: Englewood Baptist Church
Completed in 1890
Architectural firm: Bell & Swift
Geologic features: Fieldstone, Salem Limestone

Chicago boasts two remarkable Richardsonian Romanesque churches clad in Fieldstone: Edgewater's Epworth United Methodist and this little-known masterpiece. Regarding the latter, in all my years of exploring city neighborhoods no discovery has been more of a thrill, or more geologically significant. And this top-notch site is just more proof that Chicago's South Side, still an undiscovered country to all too many architecture and natural-history enthusiasts, has so much to offer. Its legacy and powerful role in making and defining this city must not be ignored.

FIGURE 11.3.   Englewood's Antioch Missionary Baptist Church is nothing less than a symphony of boulders and cobbles artfully arranged into one of Chicago's most powerful Fieldstone designs.

While an 1889 issue of the *Inland Architect* faithfully records T. N. Bell's design of "a church building, 98 by 128 feet, stone exterior" for the First Baptist Society of Englewood, neither this reference nor any other identifies the source of the exterior's stones. Most likely, though, these large and varicolored boulders and cobbles, which range from subrounded to very angular in shape, were locally collected erratics that had been dumped in our region by the late Pleistocene's Wisconsin ice sheet approximately 15 ka ago, after a trip southward of hundreds of miles. They include an interesting assortment of igneous rocks of at least Proterozoic age: pink and gray granites, and much darker mafic types as well. Before mounting on the church, each stone was split to produce a flat back side that faces inward into the mortar.

Note also the Salem Limestone ("Bedford Stone") quarried in southern Indiana that is present here in the trim and eastern-façade colonnade. Its uniform buff color and sharply defined surfaces provide the perfect, calm-and-collected contrast to the busy, bumpy, multihued Fieldstone.

AUBURN GRESHAM, SOUTH SHORE, SOUTH CHICAGO, AND PULLMAN

12.1 Dan Ryan Woods Pavilion

Dan Ryan Woods Forest Preserve (S. Western Avenue and W. 87th Street)

Designed by an unattributed government architect or design team in 1934, constructed later that year or soon thereafter

Geologic features: Camp Sag Forest Dolostone, Chicago Common Brick

The far-flung parklands of the Cook County Forest Preserve District contain a large inventory of picnic shelters, dams, wading pools, and trailside fireplaces made of our Regional Silurian Dolostone in one variety or another. This rock type, available in both flagstone form and ashlar blocks, has a warm-toned yet rugged look ideal for the rustic style that has been the hallmark of the preserve system's structures. Within this legacy of stone park architecture, the handsome Dan Ryan Woods Pavilion, constructed by the Depression-era Civil Works Administration, has special geologic significance. It's most easily accessed by walking up the path a short distance from the parking lot situated just north of the intersection of 87th and Western.

While later forest preserve structures are usually clad in Lannon Dolostone shipped in from Wisconsin, this structure is made of a more local yet rarer equivalent, the Camp Sag Forest Dolostone. This rock came from the grounds of what nowadays is another Cook County forest preserve, the Sagawau Environmental Learning Center in Lemont. In the 1930s, a contingent of that fabled jobs-creation and public-works program, the Civilian Conservation Corps (CCC), moved onto this site, dubbed it Camp Sag Forest, and took over its privately run quarry. And it was there, from the early 1930s to 1942, that CCC crews toiled to produce thousands of tons of stone—some of it containing the remains of nautiloid cephalopods and other sea creatures—in the forms of dimension stone for park projects and crushed rock for macadam pavement.

Still, the most interesting thing about the Sag Forest quarry site, which probably few if any of the CCC quarry gang ever realized, is that it stands less than half a mile to the east of a small fossilized reef, roughly 425 Ma old, exposed in a stream cut. Fortunately, that spot has also become part of the Sagawau preserve. The presence of the reef, which nowadays should be visited only in the company of a staff naturalist, suggests that the strata there are in the Racine Formation rather than the underlying Sugar Run Formation of the main Lemont quarrying district farther south. The reefal rock by the stream is edged with tilted flanking beds; farther away, as in the quarry, the strata are flat-lying, thin-bedded interreef deposits.

Here at Dan Ryan Woods, that wonderful stone forms the rock-faced, random-course ashlar of the pavilion's outer walls and piers. Note how it has the full play of earth tones characteristic of aged Silurian dolostone: cream and yellow, orange and ochre. We'd all be lucky to weather so handsomely in the passing of the years.

FIGURE 12.1. The Dan Ryan Woods Pavilion is one of many park structures in the Chicago region built of Silurian dolostone produced by Depression-era Civilian Conservation Corps crews at the Camp Sag Forest quarry in Lemont.

The pavilion's gables contain another treat: Chicago Common Brick, as pleasingly pitted and irregular-edged as ever, laid by a skilled and playful hand, with chevrons alternating with horizontal and vertical stretcher bonds. And there are signs of nature's hand here, too. More than a few of the bricks have developed white efflorescence crusts after countless cycles of wetting and drying.

12.2   St. Philip Neri Church

2126 E. 72nd Street
Completed in 1928
Architect: Joseph W. McCarthy
Geologic features: Westwood Granite, Salem Limestone, Copper

St. Philip Neri is one of Chicago's strongest statements in stone, and the crowning geological gem of its South Shore neighborhood. Though its Tudor Gothic Revival style strikes an unabashedly traditional note, there's nothing formulistic about its remarkable exterior cladding, which sets it apart from the city's other churches. Set in rugged random courses, the individual stones constitute a symphony of surfaces, from flat-faced, pockmarked, and half-spalled sections to projecting planes and bumps. And they vary in color from light bluish gray and buff to golden yellow and richer brown.

Even a professional geologist seeing this stone for the first time will find it quite a conundrum. Is it some sort of chemically altered carbonate sedimentary rock? A metaquartzite? A maddening mixture of different types? But one thing is certain at once: it's trying quite successfully to hide its identity beneath a well-developed weathering crust that makes rock identification difficult.

In fact, it's Massachusetts Westwood Granite, a Neoproterozoic igneous rock better known in the building trade as "Seam-Faced," "Plymouth," or "Weymouth" Granite. In its quarrying area southeast of Boston it is found in highly sheeted and jointed masses that resemble steeply dipping sedimentary strata. There solution-laden water has been free to seep through the cracks for untold millennia, and it has weathered the rock into its present polychromatic forms. But remarkably the Westwood has retained its solidity and structural competence, and is even marketed by its producer as "the most durable of building granites." While it's difficult to verify here, it varies in texture from an aplite (a very fine-textured granite) to medium-grained. And its mineral composition is unusual in that it contains only trace amounts of such normally typical dark constituents as hornblende and biotite. Instead, it's a fairly even mixture of quartz, plagioclase feldspar, and the alkali-feldspar microcline.

Complementing the granite is the much more familiar Salem Limestone, or "Bedford Stone." This is found in the trim, rose-window framing, and damp course. An infinitely more carvable rock type than the Westwood, the Salem here also serves as the medium for the entrance inscription and the marvelously intricate detail over the portal. And if you stand at the right angle you'll catch sight of the copper spire that rises over the intersection of the church's nave and transept. It too is a monument to aesthetically pleasing weathering since it has been allowed to make the full transition from bare-metal red to the pink and black of its oxide forms, and then on to the ultimate carbonate-sulfate patina of bright blue-green.

12.3   Drake Fountain

2935 E. 92nd Street
Completed in 1892
Architect: Richard Henry Park
Geologic feature: Baveno Granite

This Victorian Gothic fountain, a gift to the people of Chicago by John B. Drake, has the distinction of being the most peripatetic site in this book. When it was dedicated in December 1892, it stood on Washington Street across from City Hall. Then, in 1906, it was moved to a spot on LaSalle Street; three years later, to a place very far from the Loop indeed, in this small traffic triangle in the South Chicago neighborhood. Until 2020 it was also decked out with a greater-than-life-size statue of Christopher Columbus, but this effigy of the currently controversial explorer as well as two in other city parks were removed by the Mayor's Office because of fear that they would provoke political unrest or vandalism.

What remains is the fountain itself, crafted from one of the most rarely documented building stones to be found anywhere in Chicago. It's the pink Baveno Granite, quarried on a mountainside overlooking Lago Maggiore at the foot of the Italian Alps. It formed in the Permian from a rising body of magma, during southern Europe's Variscan Orogeny. This major mountain-building event, which preceded the formation of the modern Alps by about 200 Ma, was itself part of the larger story of the assembly of Pangaea.

While the fence that separates the viewer from the fountain prevents scrutiny of the Baveno with a hand lens, you'll still be able to get a good sense of its medium-grained texture and its mineral composition: pink orthoclase feldspar, white plagioclase feldspar, gray quartz, and black biotite.

12.4 Pullman Market Square

E. 112th Street and S. Champlain Avenue
Completed in 1893
Architect: Solon S. Beman
Geologic features: Porter Brick, Salem Limestone

Market Square is the hub of the Pullman community and it's the logical spot to start your exploration of this built landscape of great historic and geologic significance. The Market Hall in the center and the four curved and colonnaded apartment buildings at the corners share the same basic blend of red Porter Brick and beige Salem Limestone. While the rock type here, better known to architects as the "Indiana" or "Bedford" Limestone, is the most commonly seen in Chicago, is described in detail in various other sections, the lovely brick deserves special mention. It was manufactured just 30 miles away, in the town of Porter, in the heart of Indiana Dunes country. There, according to contemporary sources, brickmakers had been using their local Pleistocene clays since at least 1863. And by the 1880s they had the technology at hand to produce fine pressed brick, often marketed as "Indiana Red," that competed in beauty, hardness, and durability with equivalent types made in the preeminent production centers of Philadelphia and St. Louis. In all likelihood, the source clays for Porter Brick were derived not from the glacial till of the nearby Valparaiso Moraine, but instead from better sorted lake-bottom sediments that collected during a highstand when Glacial Lake Chicago, a precursor of modern Lake Michigan, was swollen with meltwater.

12.5 Greenstone United Methodist Church

11211 S. St. Lawrence Avenue
Completed in 1882
Architect: Solon S. Beman
Geologic features: Chester County Serpentinite, Berea Sandstone,
    Regional Silurian Dolostone, Copper, Lake Calumet Brick

In its manic growth and almost ceaseless expansion, late nineteenth-century Chicago was the geologic equivalent of the astronomer's black hole: it exerted an immense pull on stone producers large and small, and sucked into its confines a huge assortment of rock types from far-flung places. The urban geologist in search of the unusual will find it, and on some days may even run the risk of experiencing terminal astonishment, as I did when I first stood in front of this church and looked at its enigmatic exterior. But one thing was immediately certain to me: the Greenstone Church is clad in a green stone that isn't really *greenstone*. The latter, a metamorphosed form of basalt and gabbro, has a considerably different look to it.

The church actually sports three quite different rock types on its exterior. It's the uppermost one, set in random courses, that gives the building its name. This stone is an interesting essay in just how many shades of green there can be, from pale olive and Kelly to forest and turquoise. And it's covered in many places with white, which apparently is an exfoliation crust of calcite. This, I discovered, is one of Chicago's rarest rock types, the Chester County Serpentinite, a product of a small rural quarry south of the city of West Chester, Pennsylvania—though one prominent architectural guidebook erroneously attributes it to New England. It seems this community's founder, George Pullman, had a special liking for this stone and selected it himself for the church. So far I've found only one other (albeit less well-documented) example of its use in the city, on the Raleigh Hotel. There it's in the same heavily weathered condition.

When I contacted Mike Shymanski, the president of the Historic Pullman Foundation, I learned that local preservationists, concerned about the state of the stone, have investigated its being removed and ground up as aggregate for a new composite material that would simulate its original appearance. I do understand their concern. As the aptly named Pennsylvania geologist Ralph Stone wrote of this rock variety in 1932, "Time has shown its defects. When exposed to the atmosphere of a city it slowly disintegrates and the surface spalls." Still, I hope it can somehow be saved here in a considerably less drastic way. After all, it gives us a rare glimpse of one of the remarkable aspects of our planet's penchant for moving large chunks of itself about like pawns in the great, multimillion-year chess match of plate tectonics. The green and white rock that now graces the church began as magma that cooled to form ultramafic dunite miles below the surface, in the Earth's upper mantle. Subsequently, and probably at some point in the Lower Paleozoic era, it was plucked up, chemically altered by contact with saltwater, and resolutely pushed into the side of eastern North America at the front of another advancing land mass.

Soaring above the stone is the church's pyramidal steeple top, elegantly surfaced in copper that has weathered to a lovely and wholly appropriate light green that signals the presence of a protective patina of copper-salts compounds. Everything below it on the exterior, with the sole exception of the Lake Calumet Brick exposed on the outside of the eastern-side apse, could be thought of as a stratigraphic succession, a sequence of rock formations that each has its own story to tell. Below the Chester County Serpentinite lies a thin band of buff Berea Sandstone, uppermost Devonian in age and quarried in northern Ohio. In places it's dressed-face and in others comb-chiseled, and, like the rest of the stone, it's

FIGURE 12.2.   The northern elevation of Pullman's historic Greenstone United Methodist Church offers an interesting "stratigraphic succession" of building-stone types: from bottom to top, Regional Silurian Dolostone, buff Berea Sandstone from northern Ohio, and one of the rarest and most striking rock types seen in the city, eastern Pennsylvania's Chester County Serpentinite.

sooty in spots. It has also suffered some significant exfoliation. You can spot the Berea in other places, too—for instance, in the columns, arches, and trim.

Below the Berea lie the plinth courses of what seems here to be the least seriously spalled if still intermittently black-stained rock type. Clearly this is Regional Silurian Dolostone. Most likely it's from the Sugar Run Formation of Lemont or Joliet, though other Chicagoland quarries could have been its source instead. Here it's random-coursed and rock-faced, and it serves as the local stone component of a house of worship that is not just an architectural delight but also one of Chicago's most instructive geology lessons.

12.6   Pullman Stables

11201 Cottage Grove Avenue
Completed in 1881

Architect: Solon S. Beman
Geologic features: Porter Brick, Regional Silurian Dolostone

Set on a plinth of Regional Silurian Dolostone, the Pullman Stables offer another good look at the Porter Brick used as facing on Pullman's grander buildings.

One additional advantage of stopping by this site is that it's just across the street from another worthwhile venue, the National Park Service's Historic Pullman Visitor Center, which is staffed by rangers far more extroverted than the average Chicagoan who are eager to help you explore and better understand this neighborhood. It's also full of informative displays—including one on the locally produced Lake Calumet Brick. While you're there, you might want to ask if the nearby Pullman Administration Building, which was closed for extensive renovation at time of writing, has reopened to the public. If it has, it's eminently worth a visit.

12.7 Pullman Row Houses

Residential blocks in the vicinity of Market Square (E. 112th Street and S. Champlain Avenue)
Completed by 1884
Architect: Solon S. Beman
Geologic features: Porter Brick, Lake Calumet Brick

An exploration of the Pullman neighborhood would be sorely incomplete without a stroll up and down the residential blocks radiating from Market Square. Here is Chicago's best open-air museum devoted to the interplay between facing brick, used for façades of finer homes and larger buildings, and common brick, which usually does its duty in more obscure and humbler places.

In Pullman, the role of facing brick is played admirably by the pressed Porter Brick discussed at greater length in the Market Square section. This product of an Indiana town just 30 miles to the east is fairly uniform in its scarlet to red coloration. It graces the fronts of the row houses intended for Pullman's skilled workers, which are situated north of 113th Street.

The Lake Calumet Brick, in contrast, is the common variety virtually unique to this neighborhood. Its raw material was obtained from the ample supply of Pleistocene and Holocene lacustrine clays found at the bottom of Lake Calumet, a short distance to the east. The excavated clay was hauled to the company brickyard south of 115th Street, and there it was molded, stamped with the legend PULLMAN, and fired. Its coloration is decidedly variable; it can be anything from a pale gray to pink, grayish brown or dark brown, with the hue of each brick reflecting how close it sat to the heat source of its kiln.

South of 113th, in what originally was the area set out for the domiciles of unskilled laborers, the Lake Calumet Brick is unashamedly on display in the façades. For this application, the dark-brown variety was most commonly selected. (Personally, I think it's quite striking, especially since the dark bricks were nicely highlighted by a much lighter mortar in the joints.) It can also be found in abundance above 113th Street, especially if you, ever respectful of private property, peer down driveways and back alleys. There this common variety is exposed on countless side and rear exteriors.

THE MAGNIFICENT MILE AND STREETERVILLE

13.1 Wrigley Building

400 and 410 N. Michigan Avenue
Completed in 1924
Architectural firm: Graham, Anderson, Probst & White
Foundation: Rock caissons
Geologic features: Northwestern Terra Cotta, Kershaw Granite, Salem
  Limestone

Gleaming white and floodlit at night, the Wrigley Building has been on the short list of Chicago's greatest icons ever since its completion. The vast majority of the exterior is clad in enameled terra-cotta, cleverly graded in several different shades from the bottom's gray through middling cream tones to white at the top. Originally, this magnificent material was produced by Chicago's own giant of the trade, Northwestern Terra Cotta Works, but long after that company's demise a sizable proportion of the cladding units and their fasteners had deteriorated, and had to be replaced by identical terra-cotta sections crafted in California.

Of the rock types on display at the Michigan Avenue entrance exterior below the terra-cotta, two can be identified with certainty. Gray Salem Limestone, much more abundantly used across the boulevard at the Tribune Tower, is the damp course here, and is also present in the rusticated ashlar of the lower level facing the river. Frankly, it's not a good choice when used in direct contact with the ground, and at times I've noted white efflorescence deposits that developed due to its proximity to the plaza sidewalk and the deicer-laced water that ponds there in winter.

The other stone of note is found in the narrow pink edging strip and pink pavers below the Salem. This is the Kershaw Granite, a relatively rare sight in

MAPS 13.1 AND 13.2. Chicago's North Side and inset of the Magnificent Mile, River North, and Gold Coast neighborhoods of Chicago's North Side

MAPS 13.1 AND 13.2.   (Continued)

Chicago architecture. An attractive, coarse-grained and porphyritic igneous rock quarried in South Carolina, it contains oversized crystals of pink orthoclase, as well as smaller ones of white albite, black biotite, and clear quartz. Granites often form as a result of the plate-tectonics process of subduction, where one of the Earth's plates sinks under another, and the Kershaw is no exception. In this case, the process happened in the Carboniferous and Permian periods, as northwestern Africa approached eastern North America during the formation of Pangaea and the mighty Alleghenian mountain chain. As they sank deep into the interior, basaltic ocean crust and seabed sediments partially melted and were chemically transformed into lighter granitic magma. Less dense than the surrounding crust, this molten rock rose toward the surface. But it did not reach it before it cooled sufficiently to crystallize. Only much later did the forces of erosion expose this large body of intrusive rock, which geologists have dubbed the Liberty Hill Pluton, at the surface.

13.2   Tribune Tower

435 N. Michigan Avenue
Completed in 1925
Architectural firm: Howells & Hood
Foundation: Rock caissons
Geologic features: Salem Limestone, Mokattam Limestone, Pentelic
   Marble, Phlegrean Fields Tuff

The ultimate in over-the-top neo-Gothic skyscraper design, the Tribune Tower with its radiating cluster of nonfunctional flying buttresses is also one of the city's most massive outcrops of Salem ("Bedford," "Indiana") Limestone. This gray to buff biocalcarenite from southern Indiana's Mississippian-age Salem Formation has long been touted as the ideal architectural stone, and indeed its virtues are several. But here in recent years everything from chipped cladding panels to bad salt efflorescence and spalling were all too apparent to passersby. At time of writing a major exterior restoration project is underway as the building also undergoes an internal conversion from newspaper headquarters to yet another high-end condominium complex.

What I hope will be retained on or in the building, along with the Salem cladding old and new, is the collection of 149 rock specimens, architectural ornaments, and archaeological artifacts. These mementos from famous historical sites around the world were acquired, one way or another, by twentieth-century *Chicago Tribune* staffers—with the exception, of course, of the Moon rock provided instead by Apollo astronauts. For decades now most have been on outdoors display, embedded in the ground-floor exterior. A geologist could write an entire book, or at least an overstuffed booklet, on the scientific significance of these specimens. I here select three of my personal favorites, listed by their label texts, which with a little luck will still be accessible to the public once the renovation and conversion are complete.

"Great Pyramid, Giza, Egypt"

This chunk of gray carbonate rock is the Mokattam Limestone, quarried on the Giza Plateau and notable both for its use in the Great Pyramid complex and for its contents—spectacular fossils of *Nummulites*, a genus of the planktonic organisms known as Foraminifera. These fossils are quite gigantic, given the fact they were made by single-celled creatures. Coin-shaped and coin-sized, their shells display a remarkable unconscious artistry. While this rock specimen seems not to have any that are instantly recognizable, it's definitely fossiliferous nonetheless. The Mokattam formed in the Eocene epoch from lime mud and the remains of marine organisms deposited at the bottom of the Tethys Ocean, the Mediterranean's much larger precursor. The ancient Egyptians quarried immense amounts of this rock for use at this site, but also brought in granite and other varieties from much farther away. All in all, well over two million massive blocks of stone, with an average weight exceeding 2 tons, were required to construct the Great Pyramid.

"Mount Pentelicus Quarry; Marble Used in Parthenon"

This chaste white square of Pentelic Marble, as unblemished as any of its kind in the world, formed in the Cretaceous or Tertiary, and its rock type has been quarried since human antiquity on Mount Pentelikon, a bedrock hill some 10 miles northeast of the Acropolis. In addition to its role as the locally available ornamental and load-bearing stone of Athenian architects and sculptors, Pentelic Marble was considered a prestige item by the Romans, who exported it far and wide. It can still be found not only in the ruins of the imperial capital but in those of such colonial outposts as Leptis Magna in Libya. And much later it would even adorn the lobbies of American office buildings, all the more appreciated because of its classical credentials.

"Sibyl's Cave, Cumae, Naples"

The archaeological site of Cumae is close to Naples, but it isn't in it. Rather, it's in the *Campi Flegrei*, the Phlegrean Fields, an area to the west of that great Italian port. This is a region with a geologically violent past, a present of slumbering volcanic craters and sulfur-spewing fumaroles, and a future of perilous uncertainty due to the great magma chamber that rises and sinks fitfully beneath a human population larger than that of the city of Chicago. Legend has it that the cave dug into the hill at Cumae was home to the Cumaean Sibyl, the prophetic priestess described by the Roman poets Virgil and Ovid. The ochre-colored rock on display here is an example of Phlegrean Fields Tuff, which in turn is composed of the three igneous extrusive deposits known as Piperno stone, Campanian Ignimbrite, and Neapolitan Yellow Tuff. Which of these this specimen belongs to depends on where exactly it was collected. Regardless, each constituent of the Phlegrean Fields Tuff is the result of a separate cataclysmic ash-cloud release that occurred in the Pleistocene epoch, thousands of years before the 79 CE eruption of Mount Vesuvius sealed the fate of Pompeii and Herculaneum across the bay.

13.3   Medinah Health Club

505 N. Michigan Avenue
Currently part of Hotel InterContinental Magnificent Mile
Completed in 1929
Architect: Walter W. Ahlschlager
Foundation: Rock caissons
Geologic features: Chelmsford Granite, Salem Limestone

Like its next-door neighbor the Tribune Tower, the Medinah Club is clothed almost exclusively in beige-toned Salem Limestone. But the damp course and broad entrance trim is made of contrasting light-gray rock, the Chelmsford Granite, from Massachusetts. This is an attractive but infrequently seen stone type in Chicago. Its mineral composition is a typical one for granite: the usual lineup of muscovite and biotite micas, microcline and plagioclase feldspars, and a great deal of quartz. Still, it has a certain texture and appearance that sets it apart. It's sprinkled with both prominent black biotite specks and white feldspars that look like rice grains variously oriented lengthwise or in cross section. In places you'll see that there seems to be a hint of foliation, with light and dark crystals forming subtle bands that suggest that the rock has been somewhat metamorphosed. Here the foliation often appears more or less vertical because of the way the stone was mounted.

The Chelmsford Granite is still quarried and marketed today. Devonian in age, it comes from a pluton northwest of Boston that is part of the Merrimack Belt, a portion of the crust that apparently originally belonged to the microcontinent Ganderia, and was later profoundly affected by the subsequent collision of Avalonia, which arrived at about the time the Chelmsford magma was emplaced. Here at the Medinah Club, it must be said that the stone has taken quite a beating in places. While its carved surfaces are still sharply defined, deicer salts used in winter have caused a good deal of spalling and surface deterioration closer to grade.

13.4   McGraw-Hill Building

520 N. Michigan Avenue
Current name: The Shops at North Bridge

Completed in 1929
Architectural firm: Thielbar & Fugard
Foundation: Wooden piles
Geologic features: Salem Limestone, Morton Gneiss, Cast Iron

Though it doesn't match the height of some of its Loop counterparts, the McGraw-Hill Building is a fine example of the Grand Art Deco Formula, with a neutral mass of Salem Limestone—in this case recently removed, restored, and reinstalled—grounded by a base of a crystalline rock both somewhat darker and richer in detail. Here the latter stone is the riotously turbulent Morton Gneiss,

FIGURE 13.1. One standard version of the Grand Art Deco Formula, on display at the Magnificent Mile's McGraw-Hill Building. This structure's design incorporates a base course and grand entranceway of flamboyant and chaotically patterned Morton Gneiss from Minnesota contrasted by the unfailing uniformity of Salem Limestone everywhere above it.

the Paleoarchean migmatite discussed at length in the 333 N. Michigan section. By all means note the façade's cast iron window frames and upper-story span- drels, the magnificent carved Salem panels (four more are on display inside the building), and the big chunks of black amphibolite adrift in the Morton above the monumental doorway and in other spots, too.

13.5   Woman's Athletic Club

626 N. Michigan Avenue
Completed in 1928
Architect: Philip P. Maher
Foundation: Wooden piles
Geologic features: Larvikite Monzonite, Salem Limestone

Most of the exterior of this elegant building of Second Empire affinity sustains the Mag Mile leitmotif of Salem Limestone. But in stark contrast the ground floor is sheathed in the most opulent and extroverted of dark-toned decorative stones, Larvikite Monzonite. Here represented by its "Blue Pearl" variety, this Permian-period igneous rock quarried in southern Norway is famous for its large, labra-dorescent perthite crystals that raucously glitter away in gross defiance of the buff blandness above.

13.6   633 N. St. Clair Street

Completed in 1991
Architectural firm: Loebl, Schlossman & Hackl
Foundation: Caissons (type not specified)
Geologic feature: Lavras Gneiss

The postmodernist era of American architecture, of which this building is one of many Windy City examples, has been most gratifying to urban geologists. Build-ing stone has returned with a vengeance, and it's arrived in a multitude of forms hailing from exotic locations all around the world. While North America itself has always offered a dazzling selection of rock types suitable for all situations and color schemes, builders in recent decades have often turned instead to foreign alterna-tives. This can simply be a matter of economics: in this age of globalization, a gran-ite or marble extracted, processed, and fabricated half a planet away may actually be cheaper to transport to and install on a Chicago high-rise than one produced domestically—but one that would keep US stoneworkers gainfully employed.

Here the exotic stone is found adorning almost all of the exterior that isn't glass. What a beauty it is. Dark greenish gray with subtle wisps of dark mafic-mineral banding revealing metamorphic foliation, the Lavras Gneiss provides this structure with a somberly elegant skin. Its quarry locality, near the town of Lavras in the Brazilian state of Minas Gerais, is situated at the southern tip of the São Francisco Craton, one of South America's cores of ancient continental crust. Derived from even older, Archean granite, this striking rock type has been dated to 1.9 Ga, or about midway through the Paleoproterozoic era.

13.7   Metropolitan Water Reclamation District Building

100 E. Erie Street
Completed in 1955
Architect: Frank L. Finlayson
Geologic features: St. Cloud Area Granite, Salem Limestone

From a geological perspective there are two ways to approach this rather nondescript administrative building. You can tick off on your sighting list that it, too, is mostly faced in Salem Limestone, the way a veteran birder wearily notes the presence of yet another grackle or house sparrow. Or you can revel instead in the mineralogical complexity of the sleek and polished rock that makes up the damp course and entrance trim. This is the variant of central Minnesota's St. Cloud Area Granite marketed as "St. Cloud Gray"—here read "Very Dark Gray." It owes its unusually dark tint to its preponderance of gray plagioclase and orthoclase felspars and black biotite or hornblende. It also has its fair share of quartz, but it's clear and glassy, and does little to lighten the overall shade.

The St. Cloud Area Granite is Paleoproterozoic and dates to approximately 1.78 Ga, which means it formed a good 100 Ma after the Penokean Orogeny that affected Minnesota and northern Wisconsin. Instead, its origin may be linked to the later Yavapai Orogeny, though the role of that mountain-building event in the making of the Upper Midwest is less clearly understood than it is in the Southwest.

13.8   777 N. Michigan Avenue

Completed in 1963
Architectural firm: Shaw, Metz
Geologic feature: Lac-Saint-Jean Anorthosite

Architecturally nondescript but geologically fascinating, this building is emblematic of all the blocky high-end condo buildings of its era that feature either "black granite" cladding or a tortured twelve-ton modernist sculpture in the breezeway or, frequently, both. This particular specimen of Space Age Chic mercifully spares us the latter and rewards us with the former.

The stone ranged along its ground-floor exterior is the beautiful Lac-Saint-Jean Anorthosite. Marketed under a variety of trade names including "Canadian Black" and "Peribonka Granite," it is quarried in and near the Quebec community of Saint-Henri-de-Taillon. This intrusive igneous rock type, one of our planet's rarest, has been polished to produce a wonderful, world-reflecting mirror. By all means, pause for a moment here and watch the ghostly figures of passing shoppers flit across the alternative reality of its darkling surface.

Still, the Lac-Saint-Jean is hardly a uniform sheet of blackness. Within it swim large and lighter crystals which, if not as manically labradorescent as those of Larvikite Monzonite, do form a matrix of muted metallic glimmers. I suspect that this is the rock's magnetite or ilmenite content. Other minerals present include abundant plagioclase feldspar, olivine, pyroxene, and amphibole. Mesoproterozoic in age, the Lac-Saint-Jean Anorthosite was emplaced in the upper crust during the Grenville mountain-building phase and the formation of the ancient supercontinent of Rodinia.

13.9   Olympia Centre

161 E. Chicago Avenue
Completed in 1986
Architectural firm: Skidmore, Owings & Merrill
Geologic feature: Southern Swedish Red Granite

While one should always lament the pretentiousness of American buildings that are centres rather than centers, the fact remains that the Olympia is one of the city's most convincing examples of sheer beauty in a stone-clad skyscraper. It is also a textbook demonstration of how one igneous rock variety—in this case the Proterozoic-age Southern Swedish Red Granite—can be employed in contrasting finishes to produce impressive ornamental effects.

13.10   Chicago Water Tower and Pumping Station

806 and 811 N. Michigan Avenue
Completed in 1866 (Pumping Station) and 1869 (Water Tower)

Architect: William W. Boyington
Foundation: Shallow and spread (Water Tower)
Geologic features: Lemont-Joliet Dolostone, Valders Dolostone

Dwarfed in their setting like a pair of tugboats moored in a harbor full of ocean liners, these survivors of the Great Fire of 1871 are nevertheless the Magnificent Mile's touristic focal point. They are also the city's most famous stone structures.

The story of how and to what extent both buildings actually survived the disaster is a matter of considerable variation, with one popular account honoring a firefighter named Frank Trautman, who reputedly draped water-soaked canvas and blankets over the tower and thereby saved it from the worst of the blaze. This very likeable story makes good screenplay, but for some reason it's conspicuously absent from more scholarly accounts. In any case, as a Commission on Chicago Landmarks report makes clear, the exterior of the Water Tower, blanketed or not, got quite scorched, and the Pumping Station suffered graver damage; its wooden roof was destroyed and its interior machinery wrecked. Fortunately, however, its walls still stood, and it was ultimately restored. In both cases these structures were probably lucky to fare as well as they did. In many other places that bore the brunt of the fire, architectural stone succumbed spectacularly.

Though it was not well understood at the time, most types of rock are not fireproof. Sandstones and granites are especially unreliable because their quartz grains swell much more in intense heat than other minerals present. Limestone, dolostone, and marble may fare somewhat better if the blaze causes an outer crust of burned lime to form on their surfaces; this can at least temporarily protect the interior of the stone from further harm. On the other hand, these carbonate rocks contain a great deal of calcite, which when heated has the utter perversity to expand in one direction and shrink in another. This, too, can badly affect the rock's structural competence and promote cracking and spalling. While the vulnerability of building stone was brutally apparent in the wake of the 1871 fire, it was not until after a second major Chicago conflagration three years later that new building codes, inspired by the insistence of the insurance industry, finally required adequate fireproofing. And that came in the form of brick and terracotta sheathing. But happily for the geologists, quarrymen, and aesthetes among us, stone still remained very much in demand as ornamental cladding.

It may have a checkered history as a structural material, but who can honestly say that the Lemont-Joliet Dolostone, as seen here, is not Chicago's most beautiful rock type? Of course it lacks the dramatic patterning of Morton Gneiss. It certainly doesn't have the intricate crystalline textures of granite, the polished grace or gleam of marble, or the self-effacing workability of the Salem Limestone. No matter. It *glows*. It belongs here and it comes from here. It's intrinsically and viscerally the stone of a flat and open place informed by prairie sunsets. As its original blue-gray color weathers, as the ferrous iron within it oxidizes and hydrates, it turns that special buff to buttery tone that speaks of the land on Lake Michigan's western flank. One can revel in the almost unbelievable diversity of rock types now in this world-class city and still recognize its geologic starting point, a central and primal form of stone that is its own. And this is it.

These qualities are particularly true of the Silurian dolostone taken from the Sugar Run beds in the Lower Des Plaines River Valley, from Lemont to Joliet. Most sources suggest the rock for both of the buildings here came specifically from Joliet, but the historian Sonia Kallick, who knew that region best, states that the Water Tower's stone actually came from Lemont's Walker Quarry. The exact source of the Pumping Station's ashlar is less clear. Regardless, by the 1880s the Lemont-Joliet's faults, which were pretty much the faults of all dimension stone to one degree or another, were mostly what mattered to Chicago's architectural community. The Salem Limestone and other rock types stepped in to fill the perceived void. The local rock's lovely weathering tints were described as "tarnish"; its inability to serve as fireproofing was roundly deplored. It was the unpolishable plebian, all too common, variable in texture and color, a bit unpredictable, and not forever flawless. A great and growing metropolis was thought to need fancier stuff, a sense of grandeur, and more refinement.

Happily, today we have a more balanced view of our native rock's worth, and we should be thankful that superb examples of its obvious attributes survive here and elsewhere. The Lemont-Joliet Dolostone that has needed to be replaced here has been carefully complemented with eastern Wisconsin's Valders Dolostone, of the same age, color, and type. It promises to have a lifespan even longer than the original's.

13.11   John Hancock Center

875 N. Michigan Avenue
Now officially listed simply by its street address
Known to Chicagoans as "Big John"
Completed in 1969
Architectural firm: Skidmore, Owings & Merrill; Fazlur R. Khan, engineer
Foundation: Caissons (see discussion below)
Geologic feature: Anodized Aluminum

No description of the Magnificent Mile can ignore this mythic cross-trussed mass that casts its long and sweeping shadow on neighboring land and water. It sits as the city's lord in jaw-dropping defiance of gravity, Earth's most brutal force.

This building may be the purest human proof of the spiritual grandeur of matter, and of engineering as epic poetry.

But what would it be without its blackness? That essential alien ingredient was made possible by the cladding of anodized aluminum affixed to the external tubular-steel structure. An electrochemically treated version of our planet's most abundant metal, anodized aluminum features an oxide coating at once corrosion-resistant and decorative.

The tribulations of Big John's early foundation-laying have already been recounted in part I. Fortunately the desired stability was ultimately achieved. But when one actually stops below this ominous structure and sees it from that angle, there's the unshakable suggestion of various vectors of tipping and tilting. And the question instinctively comes to mind: *Why doesn't this monster fall over while I'm standing here?* Part of the answer lies in its impressive array of 239 caissons. Of that cluster of giant concrete pillars, 57, almost one-quarter, reach bedrock. The rest find their anchorage in the hardpan.

13.12   Fourth Presbyterian Church

866 N. Michigan Avenue
Completed in 1914
Architects: Ralph Adams Cram (main church), Howard Van Doren
   Shaw (cloisters)
Geologic feature: Salem Limestone

While it lacks any claim to geologic distinctiveness, this church is just too significant a part of its local landscape to ignore. And one can rightly appreciate this neighborhood's most Gothic expression of the ubiquitous sedimentary rock type that makes me want to rechristen the Magnificent Mile as Salem Street or, using its common name, the Grand Boulevard of Bedford. The intricate ornament around and above the church's main entrance is a testament to this Hoosier limestone's remarkable carvability. Add to that its remarkable adaptability: it's as fitting a choice here as it is on the massive verticality of the Palmolive Building one block up.

13.13   900 N. Michigan Avenue

Completed in 1989
Architectural firm: Kohn Pedersen Fox; Perkins & Will, associate
   architects
Geologic features: Massangis Limestone, Valders Dolostone, Milbank
   Area Granite, Town Mountain Granite, Carrara Marble, Vermont
   Serpentinite, Brass

No Windy City building better illustrates how traditional cladding materials returned with a vengeance during the postmodernist era. One has only to stand here, between the John Hancock Center, completed in 1969, and this 1989 building, to sense the seismic shift from steel to stone.

The predominant rock type on display on 900 N. Michigan's exterior is limestone, but for once it's not the Salem. This is apparent even at a distance, because it has a warmer aspect than the neutral-buff Indiana rock. And in the right light it suggests the yellower tones of the weathered Lemont-Joliet Dolostone on the Water Tower two blocks south. This is the "Roche Jaune" variety of the Massangis Limestone, imported from the Paris Basin of France. Jurassic in age, it has an oolitic texture—it's composed of tiny spherical grains of calcite—and it can be quite fossiliferous.

At ground level, however, you'll find two granites instead. The dark grayish-brown damp course at the bottom is the "Carnelian" brand of Milbank Area Granite, from eastern South Dakota. Neoarchean in age, this common selection is by a good measure the most ancient rock on this building. Just above it, in two different shades, is the alternately thermal- and honed-finish "Sunset Red" form of Town Mountain Granite. This product of Marble Falls, Texas, owes its origin to the extensive Grenville mountain-building event that signaled the formation of the supercontinent Rodinia at the very end of the Mesoproterozoic era. The Town Mountain is here also used for horizontal ornamental trim higher up on the façade. Unlike the Milbank below, it's porphyritic, with microcline-feldspar crystals considerably larger than the other minerals present.

The main expanse of Massangis Limestone has also been highlighted with two other rock types. On the northern and southern sides of the building, there are round white medallions of Italian Carrara Marble represented by its "Bianco Venato" variety. But a much more harmonious effect is achieved by the "Dove White" Valders Dolostone that covers the piers that run up the structure. Just a shade paler than the Massangis, it provides a pleasingly subtle contrast to the latter. About 100,000 square feet of this eastern-Wisconsin version of our Regional Silurian Dolostone was used here to complement the 330,000 square feet of the Massangis.

If you're especially eagle-eyed, you'll also notice one more stone type on 900's geologically diverse exterior. Framing the big square display windows is the veined, dark-green Vermont Serpentinite seen to better advantage at the Chicago Cultural Center. And in recent years the building's monumental main entrance has been redesigned to include a huge brass surround. This alloy of copper and zinc is a clear golden yellow in contrast to bronze's less cheerful but more august reddish-brown.

FIGURE 13.2. The massive and imposing 900 North Michigan, a high-end shoppers' emporium, grandly illustrates the late twentieth-century postmodernist return to stone—and to the concept of *contextuality*, in which new designs pay homage to a neighborhood's older ones. Here obvious effort has been made to find carbonate-rock cladding (the Massangis Limestone and Valders Dolostone) that has weathering patinas similar to the Lemont-Joliet Dolostone of the nearby Chicago Water Tower and Pumping Station.

13.14 Palmolive Building

919 N. Michigan Avenue
Now known by its address
Completed in 1929
Architectural firm: Holabird & Root
Foundation: Rock caissons
Geologic features: Addison Gabbro, Salem Limestone

As the architectural historian Robert Bruegmann has noted, the Palmolive Building is "one of the most striking skyscrapers created in twentieth-century America . . . the late 1920s skyscraper reduced to its most refined and elemental form." That said, here the Grand Art Deco Formula has been altered by a 1982 ground-level makeover. Above it, the elegant setbacks are still clad in a beige utility suit of Salem Limestone, but much of the dark-toned base is now something other than stone. Fortunately, sections of the original Addison Gabbro cladding remain on the pier bases. The composition and origin of this striking Devonian intrusive igneous rock, all too rare a sight in Chicago architecture, can be found in the discussion of the LaSalle-Wacker Building.

13.15   Drake Hotel

140 E. Walton Place
Completed in 1920
Architectural firm: Marshall & Fox
Foundation: Wooden piles
Geologic features: Deer Isle Granite, Salem Limestone

It's Italian Renaissance in style rather than Art Deco, but the illustrious Drake Hotel at the northern end of Michigan Avenue has two things in common with its neighbor just to the south. Like the Palmolive Building, most of its exterior is the architect's "Bedford Stone," the Salem Limestone. It also has a base course of Devonian igneous intrusive rock from seacoast Maine. Here, however, it's not the dark, mafic Addison Gabbro but something a few shades lighter: the strikingly coarse-grained Deer Isle Granite. Its unusual lavenderish tint is provided by its orthoclase and microcline crystals. You'll also easily recognize its white oligoclase, black biotite, and gray quartz.

13.16   Drake Tower

179 E. Lake Shore Drive
Completed in 1929
Architect: Benjamin H. Marshall
Foundation: Wooden piles
Geologic feature: St. Cloud Area Granite

After you've visited the Drake Hotel, this residential building just next door is worth a quick stop for a bit of comparative granite-spotting. After a cursory glance at its exterior it's easy to jump to the conclusion that its base course is just a continuation of the hotel's Deer Isle stone. But in fact it's another coarse-grained variety, from central Minnesota—the "Diamond Pink" brand of St. Cloud Area Granite. Close examination reveals that its large feldspar crystals are a straight-up pink in contrast to the Deer Isle's lavender-buff tone. Also, to my eye at least, there seems to be more black biotite in the matrix. But in terms of the granites' respective ages, the contrast is much more dramatic: the St. Cloud is Paleoproterozoic to the Deer Isle's relatively youthful Devonian. In taking a few steps from the Drake Hotel to the Drake Tower, you've traveled back five times farther into our planet's long history. When the Deer Isle formed, amphibians were first colonizing the land and the earliest species of trees were forming the first forests. But when the St. Cloud Area Granite solidified from its magma, there were no plants or animals at all. The highest forms of life were comparatively simple types of aquatic algae. The land itself would not turn green for well over another billion years.

RIVER NORTH

14.1 Merchandise Mart

222 Merchandise Mart Plaza
Completed in 1930
Architectural firm: Graham, Anderson, Probst & White
Foundation: Rock caissons
Geologic features: Salem Limestone, St. Cloud Area Granite, North-
western Terra Cotta, Bronze, Copper

Volumetrically breathtaking, the Merchandise Mart is best seen in its entirety from an oblique angle on Wacker Drive, across the river. There you can actually apprehend its sheer dimensionality, its monumental engineering. This immense mass stands on 458 separate caissons, concrete shafts like deep-plunging taproots reaching down a hundred feet or so to the solidity of bedrock. And from this vantage point you can also see how well the Grand Art Deco Formula works on broad-shouldered bulk as well as in the soaring setbacks of elegant skyscrapers. On this great block of a building if anywhere, the requisite Salem Limestone is not meek, bland, or merely dependable; it's glorious. This is not the human equivalent of a mountain, but of a mountain range. And note how well the Northwestern Terra Cotta ornamentation, bronze display-window frames, and roof copper complement the buff facets of rock. The latter two materials have been allowed to weather to a striking green patina.

Closer scrutiny is also called for, especially at the building's southeastern corner, where the porous Salem gives way to a damp course of the less salt-sensitive St. Cloud Area Granite. Here the exposure of igneous rock is at its deepest due to the slope descending to Wells Street. Listed in a 1935 source by its trade name of "Cold Spring Pearl Pink" and in another reference three decades later as "Diamond Pink," this coarse-grained product of central Minnesota's Paleoproterozoic granite district has long been a popular choice of Chicago architects. Its formation about 1.78 Ga appears to be linked with the Yavapai Orogeny, a mountain-building episode caused by the collision of the ancient Archean core of North America with a land mass that can be traced from Wisconsin to Arizona. While it's now part and parcel of our continent, this terrane began its journey altogether elsewhere on the globe, as an arc of volcanic islands.

14.2   353 N. Clark Street

Completed in 2009
Architectural firm: Lohan Anderson
Geologic feature: South Kawishiwi Troctolite

The landscaped seating area on the southern side of this high-rise contains planters capped with the polished version of a gorgeous, dark-gray stone of decidedly coarse texture. The same stone appears in rugged split-face form on the outer planter walls, and as considerably lighter-toned pavers. This is the South Kawishiwi Troctolite, marketed by its current producer, Coldspring, as "Mesabi Black Granite." Quarried in the heart of northeastern Minnesota's Superior National Forest and to the southeast of the town of Babbitt, it belongs to an unusual intrusive rock type that's related to gabbro, but which lacks its higher percentage of pyroxene minerals. The 353 N. Clark plaza is the best place in the city to see the South Kawishiwi in all its dark glory, and to understand how different finishes applied to a single stone type can alter its appearance significantly.

The South Kawishiwi Troctolite is composed mostly of plagioclase feldspar, with lesser amounts of olivine and augite. It formed in the upper Mesoproterozoic era as an intrusion into the upper crust, as part of the great Midcontinent Rift event that also witnessed huge outpourings of basaltic lava. At that point, about 1.1 Ga ago, the crust of what is now the Lake Superior region was torn apart, apparently by a mantle plume rising to form a hot spot, which in turn was possibly assisted by the stretching of the crust as the continent of Amazonia pulled away from Laurentia, North America's ancient precursor. Similar episodes of rifting elsewhere have resulted in the breakup of continents and the formation of new ocean basins,

FIGURE 14.1.   A plaza wall of Minnesota's South Kawishiwi Troctolite at River North's 353 N. Clark. This coarse-grained mafic relative of gabbro formed in the great Midcontinent Rift event and has been used architecturally in Chicago for the better part of a century now. When in rock-faced form, as here, it's a rugged medium gray; when dressed and polished, as elsewhere in the plaza, it's dark gray and black with glittering crystalline highlights.

but in this case the violent tectonic activity ceased before the heart of Laurentia was irreparably torn asunder. But why? An older hypothesis suggested that this occurred because another landmass collided with Laurentia, formed a zone called the Grenville Front, and exerted counterpressure. However, newer geological field data seem to indicate that the rifting had already stopped by the time the collision happened. So it may take another generation of patient research and savvy reinterpretation before we finally understand why our continent was spared.

14.3   Boyce Building

500–510 N. Dearborn Street
Completed in three phases between 1911 and 1923
Architects: D. H. Burnham & Company (original design), Christian
    Eckstorm (final design)

Foundation: Caissons of unspecified type (original section); shallow
   and spread (southern addition)
Geologic feature: Woodbury Granite

Northern New England abounds with granites that owe their existence to our continent's collision with the exotic terrane Avalonia in the Devonian period, and to the Acadian mountain-building event associated with it. And because many of these granites make excellent architectural stone, Chicago buildings are blessed with an impressive variety of them. The type on display at the Boyce Building is Vermont's gray and medium-grained Woodbury; it serves as the cladding for the first and second stories. This is stone that has been long prized for its refined and understated appearance, and it's often been chosen for such statelier settings as banks and governmental buildings (e.g., the Loop's City Hall and County Building, site 8.15). The stone panels here are mostly in good shape and show off their contents well. Look for black flecks of biotite mica, gray quartz, and three white to buff feldspars: microcline, orthoclase, and oligoclase. By the Boyce's entrance you'll also spot another, much darker stone with a riot of veins or infills of what appears to be white calcite. In architectural lingo, this intriguing rock type would definitely be called a "black marble," but unfortunately neither its trade name nor its true geologic identity have been recorded, it seems.

14.4   AT&T Switching Center

509 N. Dearborn Street
Year of completion unknown
Architect: Unknown
Geologic features: Andes Black Gabbro, Concrete, Ceramic Tile

This site provides ample proof that architectural geology can be interesting even when the architecture itself is not. No doubt designed to be virtually invisible, the utterly functional form of the AT&T Switching Center nevertheless can't help being a thing of note and beauty to the passing urban naturalist. On its western side, nestled among the concrete piers, are basal panels of Andes Black Gabbro, a Brazilian igneous intrusive rock that has been used in a number of places in the greater Chicago area. Despite its relative popularity as a "black granite" in the middle of the previous century, it has been very poorly described, and its geologic age has not been published. But a close and lingering look here reveals its large crystals of plagioclase and pyroxene. The rock also contains smaller amounts of pyrite and magnetite.

The light-blue ceramic tile also present on the western wall may not come up to the lofty artistic standards of the Rookwood lobby of the Monroe Building. But from a geologic standpoint it's as significant as the stone just described. In fact, ceramic tile could in one sense be considered stone itself. Its source material, unconsolidated clay, is composed of the microscopic mineral remains of former rock. When fired in the intense heat of the kiln, it becomes rock once more—and one fortunate enough to have a fused, weathering-resistant glaze composed of silica, aluminum oxide, and other chemical compounds derived from the Earth's crust.

14.5   Medinah Temple

600 N. Wabash Avenue
Completed in 1912
Architectural firm: Huel & Schmid
Foundation: Shallow and spread footings and piles
Geologic features: Flemish Bond Brick, Midland Terra Cotta

The Medinah Temple is a legendary location for classical-music fans. For three decades till the mid-1990s its excellent acoustics made it the preferred recording venue for the Chicago Symphony Orchestra. Built in exotic Moorish style for the Ancient Arabic Order of Nobles of the Mystic Shrine—in other words, for the Shriners—this unequivocally eye-catching building is also renowned architecturally, partly because of its amazing exterior. Brick of a striking tawny tint is laid in a distinctive Flemish Bond pattern, with alternating stretchers and headers. Note how the bricks have been textured to give the structure an ancient, exotic, and almost wind-eroded look. And the intricate terra-cotta ornament, produced by Chicago's Midland works, is also evocatively rough-surfaced.

14.6   Former Chicago Historical Society

632 N. Dearborn Street
Current name: Tao Asian Bistro
Completed in 1892
Architect: Henry Ives Cobb
Geologic feature: Athelstane Granite

On November 12, 1892, persons who'd received engraved invitations from the Chicago Historical Society's Committee of Arrangements had the honor to attend the official ceremony celebrating the laying of the datestone in the third ashlar course at the southeastern corner of this ruggedly imposing edifice. On the day following, the *Chicago Tribune* proudly reported that "The new building ... will be one of the handsomest in the city. It will be of granite, Romanesque style. .... The material used is Wisconsin rock-faced red granite for the fronts, and steel for the interior. It is designed to be as completely fireproof as possible."

As it turns out, Wisconsin was a major producer of beautiful architectural granites at the end of the nineteenth century. Determining exactly which of these is present here took me some serious igneous-petrology sleuthing, especially because the rock surfaces are quite sooty and the color and composition of their mineral content are screened by decades of accumulated grime that has given the building a considerably darker cast than it had originally. At a rock outcrop in the middle of nowhere, I'd have felt free to do what geologists have done for centuries: I'd pull out my trusty hard-rock hammer and break off a piece of the weathered stone to reveal a fresh surface. But here, I suspect, that would not be a sociable act. Fortunately, I did find, as you can, two places on the exterior where something considerably bigger and heavier than a rock hammer has already sheared off projecting portions of the granite. At these spots some new weathering has taken place, but at least the true coarse-grained texture and lighter pink hue of the stone are somewhat more apparent.

After eliminating red and pink Badger State granites from Wausau, Montello, and Waupaca I came across an obscure twentieth-century reference to this building and its "Aberdeen granite." To a British geologist, this would instantly signify the famous "Peterhead Red," quarried in Aberdeenshire, Scotland. And, as it so happens, that granite was in fact imported into the US in the late 1800s, but it was primarily used for such specialty items as ornamental polished pillar shafts on churches and graveyard monuments. But then I learned that there was also an Aberdeen quarry much closer at hand, in Wisconsin. The *Tribune*'s accuracy was vindicated.

Ernest Robertson Buckley, whose 1898 survey of Wisconsin building stones is a model dearly needed by modern geologists of how to write comprehensibly and describe fully, noted that this Aberdeen site was located in the town of Amberg. That's in the northeastern reaches of the state, almost 70 miles due north of Green Bay. There, in a region rich with granites of various tints and textures, a pink, coarse-grained stone was extracted in considerable quantities. Buckley's description seemed a perfect match for what now stands here on this exterior. But I thought it best to also check with Ken Jones, curator of the Amberg Museum, where the local granite-industry legacy still looms large. He, too, thinks this site's stone matches what came out of the Aberdeen quarry. So it's an eminently reasonable working hypothesis that the building is indeed clad in Amberg's Athelstane

Granite. That means that the rock you see here is Paleoproterozoic in age, and it hails from the time when the Superior Craton, one of ancient North America's Archean core components, collided first with a volcanic-island arc geologists called the Wausau-Pembine Terrane and then with the Marshfield Terrane microcontinent. All this crashing, crunching, and subducting produced, in addition to the lofty Penokean Mountains, a great deal of rising granitoid magma. And here you can see, some 1.86 Ga later, one of the results of this major geologic event. The Athelstane Granite, in this form, is composed of pinkish-red microcline, a lighter plagioclase feldspar, clear quartz, and black biotite.

14.7 Raleigh Hotel

650 N. Dearborn Street
Currently known by its address
Former names: The Raleigh, Vendome Hotel
Completed in 1882
Architect: L. G. Hallberg
Geologic features: Chester County Serpentinite, Regional Silurian
    Dolostone

The west side of Dearborn Street between Erie and Ontario should be proclaimed the Block of Really Rare Rock. That, or the Zone of Visually Arresting Stone. Here the Raleigh Hotel provides an almost jarring change from the pink Athelstane Granite of the former Chicago Historical Society building just down the way: the bulk of its outer cladding is a strikingly unusual olive color. This hue is indicative of a rock type that can be seen in only one other major Chicago building, the Pullman community's Greenstone United Methodist Church. While at least one architectural source correctly identifies the Raleigh's exterior as a serpentinite, the exact origin of this exotic metamorphic rock is not recorded there or apparently anywhere else. However, it so closely resembles the cladding found on the Pullman church and is so dissimilar in appearance to other stone of the same variety that it is almost certainly the Chester County Serpentinite, of eastern Pennsylvania. This stone, probably of Lower Paleozoic age, originated as upper-mantle peridotite that was bulldozed up by the front edge of a wandering land mass and then plastered onto the margin of ancient North America.

The Raleigh's English basement sports a distinctly different, cream-to-buff stone set in rock-faced ashlar. This appears to be Regional Silurian Dolostone. It was frequently used in the late nineteenth century for plinths and foundation walls. And if you look closely, you'll also see that the window sills are made of yet a third different rock type, a brownstone of uncertain source.

14.8   Ransom R. Cable House

25 E. Erie Street
Current name: Driehaus Financial Services
Completed in 1886
Architectural firm: Cobb & Frost
Geologic feature: Oneota Dolostone

The Cable House, River North's finest nineteenth-century residence, is ample proof that the Richardsonian Romanesque style can produce a stunning effect

FIGURE 14.2. The Byzantine-inspired columns and cupola carving of the Cable House entrance. This pinkish-yellow "Kasota Stone" variety of the Upper Midwest's Oneota Dolostone is one of the most striking carbonate rock types used in Windy City architecture.

even (or especially) when expressed in a lighter color. The exterior's rock-faced ashlar is the "Kasota" variety of the Oneota Dolostone, and this particular version of it has been much prized for its distinctive, pinkish-yellow tint that often makes it identifiable from two blocks away. Here it also demonstrates its competence as an excellent carving stone. Make sure you take a lingering look at the capitals of the entrance's squat columns and, above them, the magnificent riffs on the acanthus-leaf theme that curl around the base of the cupola. This building is one of the most effective uses of architectural stone in this city.

The Oneota Dolostone formed early in the Ordovician period, when its source area, southeastern Minnesota, was part of a seawater embayment between two large, upwarped structural features geologists call the Transcontinental Arch and the Wisconsin Dome. This flooding of the continental interior occurred during the Sauk Sequence, the first of the Phanerozoic eon's longstanding episodes of worldwide high sea level.

14.9   Samuel M. Nickerson House

40 E. Erie Street
Current name: Richard H. Driehaus Museum
Completed in 1883
Architectural firm: Burling & Whitehouse
Geologic feature: Berea Sandstone

The exterior of this stately residence is Chicago's great showplace of the Berea Sandstone. This buff-tinted clastic sedimentary rock has a finely gritty texture that makes it ideal for grindstones and sharpening tools, but it has also been prized as a highly workable and dependable architectural freestone. In a city that was reeling from the effects of the Great Fire of 1871, the Berea quickly gained popularity with the city's builders, and it was abundantly available because it could be expeditiously transported from its quarries west of Cleveland via the Great Lakes. However, a century and a half later this major player in the Windy City's reconstruction survives in all too few places, and then mostly as unobtrusive ornamental trim complementing other stone. But here at the Nickerson House it can be seen in its full glory, especially because it has in recent years been divested of the layer of soot and pollutants that had darkened its surface badly. Just how dismally begrimed its exterior had become is evident in the photo of the mansion in John Drury's 1941 *Old Chicago Houses*.

The Berea began its long journey to the present as sand deposits laid down in distributaries, entrenched meandering stream channels of a mighty river delta that spread in the late Devonian period where Lake Erie and Ohio's northernmost counties are today. To the east stood the lofty Acadian Mountains; to the south, the narrow bay of an epicontinental sea. North America was rotated about 45 degrees clockwise from its current orientation, and what is now the Midwest was situated in the subtropical zone south of the equator.

14.10   Episcopal Cathedral of St. James

65 E. Huron Street
Completed in 1857, only the tower survived the 1871 fire, remainder
   rebuilt in 1875
Architects: Edward J. Burling (1857), Burling & Adler (1875)
Geologic feature: Lemont-Joliet Dolostone

River North is home to what I call the Big Three LJDs, a sacred trinity of neo-Gothic churches made of luminous Lemont-Joliet Dolostone. At the St. James Cathedral, two different building phases are present in what you see today. The bulk of the structure was reconstructed four years after the historic 1871 conflagration. However, the tower and significant portions of the outer walls did survive the fire, and their stone dates from the church's original 1857 version. I've come across anecdotal accounts that suggest the soot currently staining the tower's finials is an artifact of the Great Fire, but a stereograph photo pair taken soon after the catastrophe shows them soot-free. And add to that the fact that the roof's central chimney, part of the reconstructed portion, is similarly coated with grime. The fact is that there was more than ample time in the bituminous-coal-burning decades that followed for the black film to form.

And speaking of things seemingly bituminous, you may notice that at least one ashlar block near the tower's base bears some dark residue a bit reminiscent of the asphaltum tar characteristic of the Artesian Dolostone of Chicago's West Side, but not of the rock that came from the Lemont-Joliet district. Does this isolated feature indicate the original stone came from an Artesian quarry instead? Probably not. Each source I've found indicates that all of the cathedral's masonry was indeed the LJD. This isolated example of black blobbery may be nothing more than some wayward roofing or paving sealant.

14.11   Holy Name Cathedral

735 N. State Street
Completed in 1875
Architect: Patrick C. Keeley

Foundation: Shallow and spread; four caissons of unspecified type
    added under the spire in 1915
Geologic features: Lemont-Joliet Dolostone, Salem Limestone

In *Architecture in Old Chicago*, Thomas Eddy Tallmadge wrote an interesting
composite portrait of two leading stone types in use in the early years of his own
illustrious career as an architect. The Lemont-Joliet Dolostone, he noted, was
"white, almost chalky . . . almost impossible to carve and maintaining its yellow
whiteness in a mysterious fashion against all the assaults of soot and weather that
Chicago can bring to bear." In contrast, the Salem Limestone, which he called
"Bedford stone" like everyone else of his day, "is a creamy grey or a bluish grey. . . .
It is a magnificent stone to carve but dirt sticks to it like a poor relation."

Holy Name Cathedral is the perfect place to do your own comparative petrology of these two venerable and beloved stone types. The predominant one of the
pair is the Silurian-period Lemont-Joliet Dolostone. Here, as at St. James Cathedral a few blocks away, this stone serves as the main expanse of ashlar, which is
rock-faced. However, there is also a substantial amount of trim and ornamentation made of the buff-colored, fossiliferous Salem Limestone of Mississippian
age. If it dates to the original construction of the cathedral, it constitutes my
earliest documented example of this dependable if rather anonymous rock type
on a major building in Chicago. While you may not find Tallmadge's assessment
of their relative sootiness holds up well across the city, here you have to admit
they make a good team when working together. The Salem's uniformity of tint
and the Lemont-Joliet's more idiosyncratic palette of weathering hues work in a
surprisingly beatific synergy.

14.12   John Howland Thompson House

915 N. Dearborn Street
Completed in 1888
Architectural firm: Cobb & Frost
Geologic feature: Jacobsville Sandstone

Facing the leaf-dappled oasis of Washington Square Park, the Thompson House
is a first-order geologic treasure clothed in Jacobsville Sandstone, the Michigan
version of what was generally termed "Lake Superior Brownstone." The maroon
rock here is the "Raindrop" type, quarried near Marquette, on the northern flank
of Michigan's Upper Peninsula. It was prized in its heyday for its ability to take
fine sculptural detail better than competing "Lake Superior" varieties, and also
for its distinctive darker splotches, visible here and there on the surface. These

marks do indeed suggest stone wetted by the first leavings of a passing shower. To appreciate both them and the remarkable detail of the carved ornament, look to the column shafts, capitals, and panels of the main entrance on Delaware Place—but do be respectful of this private property. This would be an appropriate site for a pair of small binoculars, but given the elegance of the neighborhood, opera glasses might be more fitting.

The Jacobsville Sandstone, which frustratingly lacks fossils and other attributes needed for precise age determination, has been assigned by different experts to various points on the geologic time line, from very late in the Meso-proterozoic to the Cambrian period over half a billion years later. (It seems that the safest bet these days is on the lower Neoproterozoic.) But whenever exactly it did form, it's clear that its sand grains were deposited after the vast outpouring of lava from the Midcontinent Rift, and were derived from the erosion of the highlands on its margin.

14.13 George B. Carpenter House

921 N. Dearborn Street
Completed in 1891
Architectural firm: Treat & Foltz
Geologic feature: Jacobsville Sandstone

This cheerfully red residence makes a nice compare-and-contrast exercise after you've taken a good look at the preceding site, the Thompson House. Believe it or not, this is Washington Square's second essay in the Jacobsville Sandstone, though it looks distinctly different. While documentary evidence is lacking, the bright brick tone of the ashlar and entrance details proclaims the fact that this is the "Portage Red" variety of the Jacobsville. It was produced about 60 miles to the northwest of the Thompson House's Marquette quarry. Until a major renovation a few years ago, this stone here presented a much darker and sootier aspect.

14.14 Harvest Bible Chapel

929 N. Dearborn Street
Former names: Unity Church, Scottish Rite Cathedral
Completed in 1867, the portion burned in the 1871 fire was rebuilt in
   1873, southern tower added in 1882

Architects: Theodore V. Wadskier (1867), Burling & Adler (1873),
    Frederick B. Townsend (1882)
Geologic feature: Lemont-Joliet Dolostone

This, the third of our River North Gothic churches constructed of rock-faced Lemont-Joliet Dolostone, makes a nice centerpiece between the stylish brownstone beauty of the Thompson and Carpenter Houses on one side and the granitic magnificence of the Newberry Library just catty-corner from it.

Like the St. James Cathedral a few blocks to the southeast, the Harvest Bible Chapel was damaged in the Great 1871 Fire, though not to the same extent. While its wooden roof perished, its walls survived. Now they bear witness to how their stone's original bluish-gray cast, a sign of the ferrous iron it contains, has been hydrated and oxidized at the surface to produce a new tint. This can best be seen from across the street at the southeast corner of the Newberry. There the subtle yellow overtones of the church's Silurian rock are most fully appreciated, both as an essential part of its architecture, really, and as a lesson in why the chemical changes wrought by the oft-feared process of weathering should sometimes be welcomed.

14.15   Newberry Library

60 W. Walton Street
Completed in 1893
Architect: Henry Ives Cobb
Foundation: Wooden piles
Geologic features: Stony Creek Gneiss, Holston Limestone, Carrara
    Marble, Shelburne Marble, Soapstone

Originally the exterior of this geologically splendid building was to be clad in the Dedham Granite, a hard, pink stone dotted with green epidote and chlorite crystals. Produced in the town of that name in eastern Massachusetts, the Dedham had been used for Boston's famous Trinity Church, but apparently not in much else of significance. As it turned out, though, delays at the source forced architect Henry Ives Cobb to cancel the contract and turn instead to the Norcross Brothers quarry in Connecticut. (The stone there, known by the trade name of "Stony Creek Granite," is actually better classified as a gneiss, albeit one originally derived from a granite.) After a visit to that site, Cobb reported to the library's trustees that he was optimistic about this being a more reliable source. Some 400 quarrymen and stonecutters were specifically assigned to the Newberry project, and, to further expedite the process, some of the Stony Creek Gneiss was first shipped to the Norcross facility in Milford, Massachusetts, for cutting before sending it on to Chicago. This fact has led at least one architectural historian to mistakenly identify the stone here as Milford Granite, which can in fact be seen at the Glessner House.

A review of Cobb's correspondence with the Norcross company, carefully preserved in the Newberry archives, reveals that not all was smooth sailing in this relationship, either. On the library's side concerns were expressed about unexpectedly delayed delivery. In turn the Norcross management, confronted with a major strike and work stoppage, shot off a lengthy explanatory telegram and separately demanded immediate payment so it could continue with the project. Happily the problems were resolved in the long run, and now the Newberry is, hands down, the most impressive granite-faced building in the city. And to this day it is, along with New York's Grand Central Station and the base of the Statue of Liberty, an example of this stone's beauty and durability proudly trumpeted by the Norcross Brothers' successor, the Stony Creek Granite Corporation.

FIGURE 14.3.   Of all the city's great cultural institutions, the Newberry Library is the most geologically rewarding. Sheathed in Connecticut's Stony Creek Gneiss, the building also boasts an interior with a variety of interesting rock types.

Most of the stone present on the Newberry exterior is in rock-faced form, but it can also be found in other finishes. The polished pillar shafts flanking the entrance are the best place to see this rock's colorful palette of minerals.

Thanks to a typewritten list of contractor instructions preserved in the construction files of the Newberry archives, we can also identify some of the striking stone originally used inside the building. For example, the rock type displayed in the entrance's vestibule, wainscoting, and step treads and risers was to be "Hawkins County Tennessee Marble, selected for handsome vein." This in fact would make it the Ordovician-age Holston Limestone, which in this guise is some of the most remarkable cladding found anywhere in Chicago. A real screamer, it's a wild mélange of brick red and gray, sliced and diced with slashes of white calcite veins. At first doubting that this could actually be a variant of the Holston, I contacted Josh Buchanan at the Tennessee Marble Company, and he in turn tracked down an old-timer employee there, who agreed that this indeed had probably been one of their earlier offerings. Later, I came across an article in a 1912 issue of the trade magazine *Stone* that specifically described the Hawkins County variant as "of the red variegated kind . . . and in general demand."

When it comes to the Newberry interior's original white and gray-veined true marbles, the archived contractor instructions simply state, with a regrettable air of nonspecificity, that they were to be selected from "the finest American or Italian marble." Still, thanks to a major restoration project undertaken from 2016 to 2018, we can at least identify the newly restored sections: Tuscany's Carrara Marble in the lobby wainscoting and countertops, and Shelburne Marble from Danby, Vermont, in the first-floor, east-side washrooms and connecting corridor.

Also in the lobby you'll spot the wall's prominent base course of black soapstone. The source of this soft but remarkably useful metamorphic rock type as installed here has not been cited, but it may be an especially dark-toned version of Virginia's Schuyler (or "Alberene") Soapstone, which is Neoproterozoic to Cambrian in age, or a black equivalent from Brazil. Then again, China, India, and other countries also export this talc-rich rock. But regardless of its locale of origin, soapstone has been prized by both ancient and modern cultures. Nowadays it's especially popular as a premium kitchen-countertop material, though it has other applications as well, including as "whiskey stones," the tippler geologist's answer to conventional ice cubes. Among its impressive properties are its low electrical and thermal conductivity, its resistance to staining and corrosion, and its workability. Though it scratches easily—please resist the urge to prove this for yourself here—marks and gouges can be removed with a quick sanding and buffing.

THE GOLD COAST AND OLD TOWN

15.1 Price House

16 W. Maple Street
Completed in 1884
Architect: Frederick B. Townsend
Geologic feature: Artesian Dolostone

Dame Fortune has smiled on this modest nineteenth-century survivor, but while it has so far dodged the wrecking ball, it is now indecorously hemmed in by newer and distinctly less handsome structures. The Price House's English basement is clad in rock-faced ashlar; that of the floors above is dressed-faced. If you look closely at the latter (small binoculars or a camera zoom lens will help), you'll see some splatters of bitumen, which identify the rock as Chicago's own Artesian Dolostone, of Silurian age.

15.2 E. Bellevue Place

Geologic features: Artesian Dolostone, Salem Limestone

With three nineteenth-century houses in a row that feature interesting building stone, East Bellevue Place is its own geological mini-tour highlighting two of the city's most historic stone types.

C. N. Fay House, 52 E. Bellevue Place (1888, Cobb & Frost)

This first stop has a front of lovely red facing brick, which unfortunately is of unknown provenance. On the other hand, its rock-faced trim and English-basement exterior have been documented as Artesian Dolostone.

Henry B. Stone House, 56 E. Bellevue Place (1887, Joseph L. Silsbee)

Here pitted and reefal Artesian Dolostone from Chicago's West Side forms the entire exterior. The bitumen staining that normally helps to identify it is harder to recognize on rock-faced ashlar, but you may spot a bit of black blobaceousness here and there.

Gerald M. Stanton House, 58 E. Bellevue Place (1888, Cobb & Frost)

Like the Stone House, this site is a Richardsonian Romanesque residence with a rock-faced façade. In this case, though, the ashlar is almost certainly the Mississippian-age Salem ("Bedford," "Indiana") Limestone. While lack of direct access prevents a check for this rock's characteristic fossil content, the evidence is strong. Its pale-gray color, which contrasts with the Stone House's golden-tinted Artesian Dolostone, and its ability to take the detailed carving present in places here, are classic Salem traits. At the time of this building's construction, this product of southern Indiana was beginning to outcompete the dolostones quarried much more locally on the city's West Side and in the Lower Des Plaines Valley. And in the years that followed, the Salem Limestone became Chicago's most common architectural stone.

15.3   60 E. Cedar Street

Completed in 1890
Architect: Curd H. Gottig
Geologic features: Murphy Marble, Copper

Chicago architects employing the Richardsonian Romanesque idiom were by no means geologically conservative. From Artesian and Oneota Dolostones to rare Montello and Athelstane Granites, they experimented widely with the excellent selection of rock types available in the city in the years following the Great 1871 Fire. In the case of this house, the "Georgia White" variety of Murphy Marble, quarried in that state's Pickens County, was chosen. Originally a Cambrian or Ordovician limestone, it was transformed later in the Paleozoic era into its present metamorphic form by the forces of crustal compression, during one of eastern North America's mountain-building episodes.

When last I checked, the striking copper trim at the top of the façade had shared some of its bright-green patina with the marble below it. Such copper-sulfate staining is an almost inevitable side effect of using this ornamentally appealing metal in conjunction with masonry.

15.4   Edwin J. Gardiner House

1345 N. Astor Street
Completed in 1887
Architectural firm: Treat & Foltz
Geologic feature: Dunreath Sandstone

In the highly unlikely event Chicago had a Department of Geologic Conservation and I were its commissioner, I'd ask the mayor to cordon off this residence, declare it a Grade A Protected Urban Outcrop, and petition the federal government to make it a national monument. That's because the rock that adorns the place is more than a rarity, it's a gorgeous rarity known as the Dunreath Sandstone, from Marion County, Iowa. I suspect it has been used elsewhere in the city, either in the polychromatic form on display here, or in its more consistently red variety. But so far I've found no evidence of it.

The architect William B. Lord, writing of the Dunreath in the February 1887 issue of the *Inland Architect*, noted it was "a most singular mottled sandstone,— red, white and bright yellow. . . . It is certainly a very odd stone and easily worked." This less-than-ringing endorsement did not keep the Dunreath Red Stone and Quarry Company from singing its praises in an ad in the same journal at about the same time:

> OUR VARIEGATED STONE is a unique and thoroughly artistic natural material, and Architects desiring a new article, and in search of new effects, will find in it an opportunity of introducing into city and suburban residences the variegated rustic work which forms such an interesting feature of European residences.
>
> Being a pure Silica, and free from clays and lime, this stone is not affected by the weather, and does not bleach on exposure.

Despite these clear references to this stone's Iowa origins, there is some modern confusion about it. One prominent architectural guidebook incorrectly states

FIGURE 15.1.   A rock-faced rarity. The Gold Coast's Gardiner House is clad in the polychromatic and crossbedded Dunreath Sandstone, which was quarried in central Iowa for a brief time.

that it was produced in Ohio. This led me to check with some of my contacts in that state, and one, Kevin Aune at Briar Hill Stone, gave me an explanation that probably was the source of the misconception: he'd found out that three Ohioans had migrated to Iowa in the 1880s and started the Dunreath operation there. I also learned from veteran Illinois geologist John Nelson, an expert in Pennsylvanian stratigraphy who knows the geology of the Dunreath area well, that both the quarry and its town are now resting under the sparkling surface of Lake Red Rock. This manmade body of water, intended to control regional flooding and provide Iowans with a new recreational resource, was created in 1969 by the damming of the Des Moines River.

The Dunreath Sandstone was taken from the lower portion of the stratigraphic section known as the Cherokee Group, and therefore dates from the Pennsylvanian subperiod of the Carboniferous. This was a time when North America was part of our planet's most recent supercontinent, Pangaea. The low-lying Midwest, straddling the equator, experienced cycles of inundation and reemergence due to the rising and falling sea level, which in turn resulted from episodes of glaciation and melting in the Southern Hemisphere. The region was home to an ancient version of tropical rainforest replete with giant insects, millipedes, and amphibians, and it was from the abundant vegetation in which they dwelled that our nation's most widespread coal deposits ultimately formed. Because the Gardiner House's stone seems to have signs of crossbedding, it's likely that its sand was originally deposited by flowing water in some long-vanished Coal Age stream channel incised into the underlying Mississippian strata.

15.5   George A. Weiss House

1428 N. State Parkway
Completed in 1887
Architect: Harald M. Hansen
Geologic features: Murphy Marble, Copper

What was noted for the 60 E. Cedar Street residence is also applicable to this Richardsonian Romanesque house, except that here the exterior stone is the "Etowah" brand of Murphy Marble. It's the pink variety that shares the same quarrying region and Paleozoic origin with the other house's "Georgia White." An 1887 *Inland Architect* description offered high praise for both the building and its rock type: "We venture to say this is one of the finest appearing stone façades in Chicago. The elegant carvings, notwithstanding its hardness, that adorn the façade, show beyond dispute that the stone works kindly under the chisel." And how well the weathered green copper trim complements the rose-tinted masonry!

15.6   Russell House

1444 N. Astor Street
Completed in 1929
Architectural firm: Holabird & Root
Geologic feature: Bois des Lens Limestone

Providing a distinct contrast to the Gold Coast's abundant crop of Romanesque revival mansions, the Russell House is, in its vertical emphasis and sleek styling, quintessentially Art Deco. And it features one of Chicago's rarer rock types, the Bois des Lens Limestone. This almost shockingly white carbonate hails from Moulézan, in the Languedoc-Roussillon province of southern France. It has a very long record of use stretching back to the fourth century BCE. In their day ancient Roman builders made good use of it, most famously for *La Maison*

Carrée, the lovely Corinthian-order temple in Nimes that inspired Thomas Jefferson and many other architects as well.

Petrologically speaking, the Bois des Lens is one of the Mediterranean Basin's endless array of Mesozoic limestones associated with the carbonate platforms of the great Tethys Ocean realm. Specifically early Cretaceous in age, the Bois des Lens is fine-grained and oolitic in texture, which means that it's composed of tiny spherical calcite granules. This fact suggests that it originally formed in shallow water well agitated by wave or tidal action.

15.7   1500 N. Astor Street

Former names: Patterson House, Cyrus McCormick Mansion
Completed in 1893, northern addition in 1927
Architects: McKim, Mead and White (1892), David Adler (1927)
Geologic features: Roman Brick, Terra-Cotta, unidentified brownstone

A Renaissance-style palazzo designed by one of the most eminent New York City architectural firms of its day, this handsome building offers an interesting and effective mixture of colors and materials. The exterior's ground floor is clad in a maroon brownstone whose exact identity remains a matter of mystery. That said, it works nicely as a suitably somber underlayer for the pastel-orange Roman Brick of the upper floors. Also note how well the longer and flatter shape of the bricks adds to the visual appeal, as does the terra-cotta trim and ornamental details.

15.8   George E. Rickcords House

1500 N. Dearborn Avenue
Completed in 1889
Architect: William W. Clay
Geologic feature: Montello Granite

Of the great influx of exotic rock types available in the era when this house was built, architect John Wellborn Root wrote, "Not a day passes but specimens are brought of new materials, granite from Wisconsin, sand-stones from Michigan, onyx from Mexico, marbles from Colorado to California. . . . With a wholesome quality of mind and life in the layman and with imagination and discrimination in the architect, what may not our domestic architecture become? In twenty years this will be the richest and most luxurious country ever known on the globe."

One such new introduction was Wisconsin's beautiful and remarkably tough Montello Granite. In the decades following its arrival on the scene it was a popular selection in the monumental trade—the business of making and marketing tombstones, grave markers, and monuments. It also was widely used as pier riprap and as pavers in the days before the widespread introduction of asphalt and other modern road-surfacing materials. Its role as an architectural stone, however, was always less extensive, certainly not because of its lack of attractiveness or durability, but because it's almost supernaturally hard. And that equates to its being difficult and time-consuming to cut, finish, and carve. In 1898, the geologist Ernest Robertson Buckley listed five buildings in this city, including this one, that were built with it; the other four have succumbed to what could be politely termed neighborhood mutability. Tantalizingly, Buckley also noted that there'd been "many others" as well—but if that's so, their identity is lost. As it stands, the Rickcords House is the one surviving documented example.

To best understand the nature of an architectural stone, one has to travel to its source and see the site and region from which it came. If you're willing to make a geological pilgrimage in the form of a seven-hour round-trip drive from Chicago, you'll find the small town of Montello in east-central Wisconsin well worth a visit. In its heart is the quarry that produced the Rickcords stone; it has now been turned into a park and pond replete with a wishing well and waterfalls. There you can directly inspect outcrops of the granite. When this quarry was in operation, from about 1880 to 1976, its crews removed the rock sometimes by blasting with black powder or an explosive called coalite, but most frequently by drilling 4-inch-deep holes spaced every 5 inches or so, and inserting wedges into the holes that were hammered till the hard rock cracked, usually to the depth of 8 to 10 feet. Later, the hand-drilling was replaced with channeling machines using mechanical drills or high-temperature flame jets. But because of the prevalence of unsightly quartz and hematite veins, the majority of the removed rock was unusable, resulting in an alarming wastage rate of about 90 percent. That, too, did nothing to make the Montello more economically competitive. And as Bryan Troost, the quarry's last owner, has told me, conditions for those working on-site were challenging by anyone's standards. Besides the danger of contracting silicosis, the bane of granite quarrymen everywhere, long hours spent in the great hole meant exposure to sun, dust, and extreme temperatures. It could reach 120°F in the summer, and -20°F in the depths of winter. But the work went on regardless.

Once removed from the headwall, the stone sections that were deemed acceptable were cut nearby with saws with giant iron teeth and steel shot or, from the 1940s onward, with braided steel wire. Using that technology, most granites can be sliced through at a rate of about 4 feet per hour. But with the unusually resistant Montello, only 8 inches per hour were possible, and the cutting wires had an especially short lifetime of only two days. These factors contributed to this stone's increased cost. Pieces to be polished, such as the squat pillar shafts on the southern side of the Rickcords House, were then placed on a cast iron polishing wheel, where they were abraded and smoothed with either steel shot or a slurry of water and emery, an abrasive powder made mainly of the mineral corundum.

The reasons for the Montello's exceptional strength and hardness lie in its origins about 1.76 Ga ago, in the Paleoproterozoic era. While most of the architectural granites on display in Chicago are the products of crustal thickening due to converging plates, subduction, and mountain building, the Montello rock formed from magma produced when there wasn't any such activity in that part of the Upper Midwest. Recently, some specialists have hypothesized that the crust was actually stretching, thinning, and even rifting in places instead. In any case,

FIGURE 15.2. Devilishly hard to cut and dress, Wisconsin's quartz-saturated Montello Granite was in the heyday of its quarrying most often used for enduring cemetery monuments and virtually indestructible road pavers. But the Gold Coast's Rickcords House is an impressive instance of its rarer role as building stone.

this rock's mineral content is unusual, and unusually simple. Classed as an alkali-feldspar granite, the Montello lacks the plagioclase found in many other granites but instead has a superabundance of red orthoclase as its one feldspar constituent. The only other mineral present in any quantity—and it makes up about 40 percent of the whole rock—is gray quartz, which accounts for the stone's great solidity and durability. While the magma that became the Montello Granite did not reach the surface before it cooled, much of the rest of that molten-rock body did, and it triggered an immense volcanic eruption that blanketed the region in hot ash, which in turn became the rhyolite that can still be found outcropping in various locations in central Wisconsin.

At the Rickcords House, most of the Montello Granite is in rock-faced form, in good Richardsonian Romanesque fashion. But there's also a small sampling of carved stone, no doubt the work of a mason with strong wrists and the patience of Job. Here, the stone has an elegant, cool-purple color, but elsewhere it can be much closer to cherry red.

15.9   235 W. Eugenie Street

Completed in 1962
Architectural firm: Harry Weese & Associates
Geologic feature: Chicago Common Brick

> Common brick . . . always means the commonest—and the bricks of Chicago are not of good renown. (*American Architect*, 1877)

While this quotation reflects a prevailing East Coast view of the quality of Chicago's own resources, Chicago Common Brick has always had its own hometown defenders. As F. L. Hopley wrote in a 1901 *Chicago Examiner* article, "Throughout the world Chicago is known as the 'Windy City,' but a much more correct appellation would be 'The City of Brick.' Like some of the more humble mammals, Chicagoans have literally builded their homes from the mud on which they have taken up their habitation." He then presents a compelling image of a metropolis derived from its substrate of clay: "For mile after mile there are crowded on either side only buildings of brick—brick in every direction—acres and acres of brick. It makes little difference as to what the front of a house or business block is faced with, whether it be brown stone or some fancy brick imported from other sections of the country; but the building itself—its walls and partitions—are constructed of the building brick manufactured right here in Chicago. No city in the world, not even New York, has found such general use for common brick as we have found here on Lake Michigan." Hopley also

FIGURE 15.3.   Be it ever so humble, there's nothing quite as trendy nowadays as Chicago Common Brick. A stalwart of the Windy City building trade in decades past, it's no longer produced, but it is rampantly recycled. This façade pier at the 235 West Eugenie Street apartment building demonstrates that the Chicago Common has emerged from its former secluded haunts to take on a new role as a pebbly, pitted, and salt-encrusted celebration of the Natural Look.

proudly noted that the annual output of Chicago's brickyards at time of writing would fill 67,000 freight-train cars; were the bricks stretched end-to-end, they'd reach halfway to the Moon.

By the time this Old Town apartment building was constructed, Chicago Common, this soot-stained Cinderella of the brick trade, was shedding its original humble role. No longer relegated to side and rear building elevations, or to places indoors not intended for the public gaze, its earthy attributes—the variable coloration and the imprecise shapes—were becoming overtly fashionable.

Perhaps, in this time of the rise of the American environmental movement, it was somehow thought closer to nature, or to what nature intended clay to be. In any event, by the time the last Chicago Common producer closed in 1981, this soft, pebbly, and pitted brick had long been an item of architects' desire. Nowadays, the recycling businesses that scavenge it from demolition sites and resell it are its sole remaining source for new projects.

Here on the external piers and walls the Chicago Common displays its classic traits. Peach to salmon to a richer red, and often coated with a white efflorescence bloom, it here and there still has surviving bits of our native Silurian dolostone that signal that its source clay was dug from the Pleistocene glacial till that blankets much of our region. In some of the bricks there are also interesting oval and ringed patterns that probably reveal the migration of salt compounds through the porous interior during repeated episodes of wetting and evaporation.

15.10   Midwest Buddhist Temple

435 W. Menomonee Street
Completed in 1972
Architect: Hideaki Arao
Geologic feature: Concrete

Despite its long history of use in some of architecture's most famous structures, concrete is usually considered a drab, workaday substance worth only a quick nod of appreciation for its prevalence and amazing utility. But at the entrance of the Midwest Buddhist Temple its significance as an ornamental material is more dramatically apparent than anywhere else in the city. The wall flanking the gate is composed of a very coarse-textured concrete with an aggregate of both rounded and angular pebbles. This surface has been vertically and deeply scored, and the projecting portions have been bush-hammered so that each has its distinctive topography: a pleasing play of variation within sameness. From a few yards away the effect resembles a medieval farmer's ridge-and-furrow field as seen from above.

This serene oasis in the urban hubbub is the quintessential place to contemplate the inner nature of concrete, which is really as much of a rock as any to be found at a quarry or outcrop. It just happens to be the product of the geologic process known as human meddling. (All our wishes to the contrary, we're as much an instrument of nature as anything else.) But the temple's concrete refrains from participating in the old cycle of natural limestone burned to produce lime, which in turn is used to create manmade limestone. Here a distinctly different rock type has come into being. Cognizant of my duty to further science's sacred

FIGURE 15.4.   Old Town's Midwest Buddhist Temple, with its artfully furrowed concrete, reveals one aspect of this workaday material's ornamental potential.

mission of creating as many pretentious and multisyllabic terms as possible, I hereby reclassify it calcareous anthrobreccia.

15.11   Ulysses S. Grant Memorial

Cannon Drive just south of Lincoln Park Zoo
Completed in 1891
Architect: Francis M. Whitehouse
Sculptor: Louis T. Rebisso
Geologic features: Vinalhaven Granite, Bronze

Chicago's bronze statuary usually comes in one of two predictable colors, deep brown and bright green. Both are products of this alloy's copper content eagerly reacting with its environment. The dark patina, as seen in this equestrian portrayal of the great Civil War hero and eighteenth president of the US, is produced artificially in the artist's studio with a special chemical treatment that creates a more-or-less permanent oxide coating at once somberly attractive and protective against further corrosion.

The statue, forged by the Ames Manufacturing Company of Chicopee, Massachusetts, is mounted atop a grandiose arched and rock-faced Romanesque base that certainly does not speak to Grant's famous lack of pretension. However, the stone itself, the Vinalhaven Granite, might be taken as a valid allusion to the general's stolid and implacable command presence in battle. Also known as "Fox Island granite," the Vinalhaven is one of several igneous architectural-stone selections taken from the Ganderia Terrane of Maine's seacoast. Silurian in age, it's gray in overall cast. This monument provides the urban geologist with a superb and extensive exposure of its coarse-grained mineralogy, which features buff-tinted orthoclase (the alkali-feldspar component), black biotite, white oligoclase (the plagioclase feldspar) and gray quartz.

LOGAN SQUARE, LINCOLN PARK, AND LAKE VIEW

16.1 Illinois Centennial Monument

Logan Square (N. Milwaukee Avenue and W. Logan Boulevard)
Completed in 1918
Architect: Henry Bacon
Sculptor: Evelyn Beatrice Longman
Geologic feature: Holston Limestone

The Holston Limestone, better known in the building trades as "Tennessee Marble," is a frequent sight indoors, where in its several color variants it does good duty as floor pavers and hallway wainscoting, as seen, for example, at the Chicago Cultural Center and the Newberry Library. At the Illinois Centennial Monument, however, it plays a more daring role, as a calcareous rock type completely exposed to the elements. The version employed here is its lovely pink variety, and fortunately it and sculptor Longman's graceful and dignified bas-relief figures have held up well over the course of a century.

The Holston Limestone dates to the Middle Ordovician, when it formed from lime muds that accumulated in a shallow marine environment off the coast of Laurentia, North America's Paleozoic predecessor. Later in the same period these deposits were pushed farther onto the continent and deformed by an advancing arc of volcanic islands. The resulting tectonic stresses probably account for the Holston's one most diagnostic feature, which can be seen in abundance here: the dark-gray, crinkled, and wavy lines known as stylolites. These lines actually represent jagged surfaces that run through the stone. Geologists think these features can form even in undisturbed sedimentary rock when, under stress from the weight of overlying strata, it experiences the migration of films of soluble ions within it. Clays and other insoluble substances remain, however, and make the stylolites visible. But in the case of the Holston they're more likely a byproduct of the pressures generated by the collision of the two landmasses some 450 million years ago.

One additional point of interest here is the monument's basal platform, which is made of an extremely coarse-textured pink granite that, regretfully, is of

FIGURE 16.1. The Holston Limestone, better known as "Tennessee Marble," has been used for flooring and wainscoting in countless Chicagoland interiors. But at Logan Square it's on view outdoors at the Illinois Centennial Monument. Here it demonstrates its often overlooked potential as an excellent carving stone.

unknown affinity. Its hefty feldspar crystals indicate that its source magma took a very long time to cool and fully solidify underground.

16.2   Logan Boulevard Graystone District

W. Logan Boulevard East of Logan Square
Architects: Various
Geologic feature: Salem Limestone

While the homes that line this leafy boulevard on either side are hardly monolithic in the truest sense of that word, there are a significant number of graystones—houses with exteriors of that most common and versatile of American architectural rock types, the cement-gray Salem Limestone—the "Indiana" or "Bedford" Limestone of the building trades. You'll spot them sporting various architectural styles, scattered among their companions faced in red brick and other materials instead. (My own favorite, with its rock-ribbed porch, piers, balustrade, and balcony, is on the southern side of the boulevard, at 2819 W. Logan.) This neighborhood is yet another demonstration of how, by the end of the nineteenth century, this product of southern Indiana had come to dominate Chicago's stone trade.

16.3   Theatre School at DePaul University

2350 N. Racine Avenue
Completed in 2013
Architectural firm: Pelli Clarke Pelli
Geologic features: Buddusò Granite, Antalya Limestone

This building may suggest nothing quite so much as a shipping container that's landed on a stack of flat cardboard boxes, but it features two interesting Old World rock types, one igneous and the other sedimentary. Starting at ground level you'll find a damp course and planters of the "Luna Pearl" variety of the coarse-grained Buddusò Granite, from the rugged Mediterranean island of Sardinia. This intrusive rock dates to the late Carboniferous or early Permian and comes from the great Corsican-Sardinian Batholith that formed during the Variscan mountain-building episode. It's been further parsed by European geologists as both a monzogranite (a granite with a feldspar content that's from 35 to 65 percent plagioclase) and a leucogranite (a light-hued granite whose magma was derived from the melting of metamorphosed sedimentary rock units in the higher portions of the Earth's crust during continental collision). Its suite of minerals includes both alkali feldspars and the plagioclase already mentioned, as well as biotite and amphibole.

Above the Buddusò and forming the bulk of the cladding is the white Antalya Limestone, from the Finike district of Anatolia, Turkey. Also marketed as "Limra Limestone," this carbonate rock, which a recent Turkish Geological Survey map indicates is upper Cretaceous in age, is available in different tints. The stone here is remarkable for its chaste whiteness.

16.4   James P. Sherlock House

845 Belden Avenue
Completed in 1895
Architect: Louis Brodhag
Geologic feature: Jacobsville Sandstone

A particularly pleasing example of the Richardsonian Romanesque, the Sherlock home requires no special sleuthing to uncover the true identity of its exterior stone. The chief clues, apparent without magnifying glass even to those of modest detective skills, are the rock's rich maroon color and its sporadic dappling of darker spots. These solve whatever mystery this open-and-shut case might have potentially held for the investigator. As with River North's Thompson House, this is the Jacobsville Sandstone, or "Lake Superior Brownstone," in its highly prized "Raindrop" variety from Marquette, Michigan.

On residences of this architectural style, column shafts were often separately fabricated from polished red granite instead. Here, however, the same maroon sandstone with a comb-chiseled finish has been used.

16.5   Lincoln Park Presbyterian Church

600 W. Fullerton Parkway
Former name: Fullerton Avenue Presbyterian Church
Completed in 1888
Architect: John S. Woollacott
Geologic feature: Napoleon Sandstone

In the architectural literature of the late 1800s there are various references to a mysterious "Michigan Buff Sandstone," also described as "Michigan Buffstone" and "Michigan Green Buff." Buildings that featured it are cited; its current prices are listed. But where exactly in that state it came from was for some reason never mentioned.

One surviving Chicago building that does display this intriguing rock type in ample quantity is the Lincoln Park Presbyterian Church, where it forms the rock-faced exterior so beloved of architects employing the Romanesque idiom. This place is of decidedly eccentric mien. For one thing, its strangely hued and random-coursed ashlar gives it a rather rough-and-tumble look, which is further heightened by buttresses of stacked discoid blocks. These no doubt are solidly secured but they still seem liable to teeter and collapse at any moment. All in all this is a site that's bound to draw passing geologists to it like wasps to a picnic jam pot.

After eliminating every other known possibility, I've decided that this distinctive rock with a slightly olive tint is the Napoleon Sandstone, from the Lower Peninsula's Jackson County and neighboring areas. The upper member of the Mississippian Marshall Sandstone Formation, it formed in a shallow, wave- and current-swept marine environment near what was then the shoreline of a large epicontinental sea. The stone's greenish cast is due to the presence of the mineral chlorite. Intriguingly, it also contains interesting sedimentary structures, some more than 2 feet across, that appear to represent sand deposits that filled holes created by tidal whirlpools. Termed rippled toroids, they're similar in size and shape to the church's buttress blocks, and it's easy to wonder if that is what the stonemasons applied a rock-faced finish to and then used here. However, the lower sections of stone appear to be rounded only on their projecting sides. This suggests that they were fashioned from normal blocks instead. However, at least some of the smaller stones near the tops of the stacks do more closely resemble actual toroids and could conceivably be examples of this very rare geologic feature.

16.6   Reebie Storage Warehouse

2325–2333 N. Clark Street
Completed in 1922
Architect: George S. Kingsley
Geologic feature: Northwestern Terra Cotta

Ankh if you like terra-cotta! Pharaonic magnificence is usually not associated with commercial storage warehouses, but here on North Clark it manifests itself in polychromatic burnt clay crafted about twenty blocks westward by the Northwestern Terra Cotta Company. May the gods of Egypt bless both architect Kingsley and original owners William and John Reebie for thinking up this delightfully incongruous addition to the Lincoln Park neighborhood. Intent and grim-faced Chicago needs more incongruity.

FIGURE 16.2. Chicago's North Side is home to many delightful examples of fanciful terra-cotta art, including this unexpectedly pharaonic version of a storage warehouse. The ornament here, expertly executed by the Northwestern works, includes two door-flanking, olive-green Ramses II statues. Note also the various botanical references on the columns.

Restored and cleaned in 2016, the exterior is in excellent shape and constitutes one of the city's greatest monuments to its terra-cotta legacy. Twin olive-skinned figures of Ramses II, inscribed below with the Reebie brothers' first names in hieroglyphics, guard the main entrance. The botanically inspired urban naturalist will also find palm, papyrus, and lotus motifs; those trending toward astronomy, the winged god of the midday sun. But however ancient these Egyptian themes may seem, they pale in temporal significance to the raw material used to make this terra-cotta. It was mined from shales and underclays of the Illinois Basin and then shipped by rail to the Northwestern works on the eastern bank of the Chicago River. These fine sediments had formed the shallow sea bottoms and coal-swamp soils of the Pennsylvanian, some 300 Ma ago.

16.7 Elks National Memorial

2750 N. Lakeview Avenue
Completed in 1926

Architect: Egerton Swartwout

Geologic features: Exterior Salem Limestone, Bronze; interior featuring a large assortment of ornamental stone types discussed below

*A pilgrimage to this geologic jaw-dropper is an utter necessity for any person who
seeks to understand the beauty and diversity of Chicago's ornamental stone varieties and what can artfully be done with them. Rockhounds who pay a visit should
be accompanied by a responsible adult alert to the early warning signs of giddiness,
euphoria, and hyperventilation. Here, given the staggering assortment of different
rock types, my emphasis must necessarily be on sketching out that diversity rather
than delving into greater detail for particular stone varieties. However, their geologic ages will be indicated, when known.*

Serving as the headquarters of the Benevolent and Protective Order of Elks
and as a memorial for its members who died serving this country in war, the
building is the best marriage of "Bedford Stone" and the Beaux Arts style Chicago has to offer. And that's quite an achievement, given its competition. More
accurately known as the Salem Limestone, this rock type is Indiana's great gift
to this city and the rest of the nation. Here it presents an elegantly uniform look
that's a dignified contrast to the cacophony of flamboyant rock types contained
within. Its reputation as a first-rate carving stone is upheld here by the still-crisp
details of the encircling bas-relief and inscriptions, now almost a century old.
Also of note are the two bronze elk statues that repose on either side of the base
of the entrance steps. Both are suitably clothed in copper-compound patinas
that protect them from destructive weathering in the acid- and pollutant-laced
urban air.

Indoors, the panoply of ornamental stone both foreign and domestic is
almost too much for the human nervous system to take in. And it makes for a
daunting geologic research project, as I can personally attest. Surviving accounts
and in-house records constitute a labyrinth of sometimes contradictory information that cites long-obsolete trade names, not infrequently misspelled or mis-
interpreted, that seem to have left no trace in the literature. Despite that, I've
learned it's possible to piece together the identities and locations of most of the
rock types here.

Beginning in the entrance area, two stone types are most apparent. The white
and dark-veined wall cladding here and in much of the rest of the building is
the "Eastman's Cream" variety of Ordovician-age Shelburne Marble, quarried in
West Rutland, Vermont. Some 22,000 cubic feet of this stone were required for
the job. Also in the entryway are two huge urns set in niches. They and their
stands were fabricated from Triassic-period Portoro Limestone, also known as
"Black and Gold Marble." The dark matrix of this carbonate from the vicinity of

La Spezia, Italy, is due to its bitumen and sulfide-compound content, which is an indicator that the rock formed from sediments that accumulated in the biological desert of an anoxic seafloor.

The floor of the rotunda's lower level contains a magnificent expanse of inlays set in a matrix of pinkish-brown "Tennessee Marble"—the Ordovician Holston Limestone. Strips and inlay borders consist of gray, fossiliferous, and in places stylolitic Warsaw Limestone (Mississippian), marketed as "Carthage Marble" in honor of the Missouri town that produces it. The inlays themselves offer a dramatic contrast of choice European stone types, which include the deep-green

FIGURE 16.3. A section of the Elks National Memorial rotunda. From gleaming white walls of Vermont's Shelburne Marble to columns and floor insets of rare limestones, breccias, ophicalcites, and true marbles, this building offers the urban geologist an exhilarating foray into sheer sensory overload.

Larissa Ophicalcite from Thessaly, Greece (Jurassic or Cretaceous); the red and light-veined Levanto Ophicalcite from Italy (probably Jurassic); southern France's black and pale-green-veined Maurin Ophicalcite (Cretaceous); and, also from southern France, the Devonian Languedoc Limestone, with a network of white or pale-blue calcite veins set in a matrix of bright red.

A quite different selection of rock types was chosen for the lower-level columns supporting the circle of arches and in the outer ring as well. They include different varieties of Italy's Seravezza Breccia (Tertiary), the Greek Skyros Breccia, the "Eastman's Cipollino" version of Shelburne Marble, and the "Madre Veined Alabama" brand of Sylacauga Marble (probably Cambrian or Ordovician). The Seravezza examples, quarried only a few miles away from the famed marble center of Carrara, are distinguished by large angular white clasts set between narrow channels of violet or dark gray. The Skyros, in my mind the most striking of all, has rock fragments of white, tan, and darker brown floating weightlessly in a matrix of inky blackness. And "Eastman's Cipollino" is a nonbrecciated marble with sinuous waves of green and deep-gray; the Sylacauga is a chaste white with sporadic black veining.

The columns of the rotunda's upper level are, once again, all different choices from those below them, though some of their stone was also used in the lower-level floor inlays already discussed. This petrologic abundance demonstrates what a wide array of rock types that the building trades define as "marbles" were available in the 1920s, and in some cases still are. Here the architect seems to have been on an "I'll take three of each" shopping spree that in hindsight can be regarded, geologically at least, as a wonderfully sound investment. The tally of rock types for this set of column shafts:

- *Five green and sometimes also deep-red ophicalcites*: the Maurin discussed above; the "Alps Green" variety, which may refer either to the Aosta Valley or Polcevera types, both from northern Italy; another Italian stone, the red Levanto, cited as an inlay above; and the Tinos (Cretaceous) and the Larissa (Jurassic-Cretaceous), both from Greece.
- *One breccia*: the French Pourcieux, marketed as "Jaspé du Var," which is pink-red with abundant white veining.
- *Two red limestones*: the Languedoc, already mentioned as an inlay, and the "Rouge de Rance" variety of Belgian Red Limestone (Devonian, fossiliferous, with white calcite fillings).
- *Two persistently mysterious selections*: a stone cited in a contemporary article as "Rouge Rubo" and in the Elks records as "Rubo (Austria)," which may be a forgotten or mangled trade name for the lovely red Adnet Limestone quarried in that country; and the even more perplexing

"Sonora (French Alps)," apparently a limestone, which is red with white and blue streaking and seems very similar to the Languedoc also from that general region.

The corridors leading off the lower-level rotunda are also worth a visit, because they're floored in another Mississippian-age carbonate rock, the Burlington-Keokuk Limestone, quarried in Phenix, Missouri. Marketed as "Napoleon Gray Marble," this selection contains easy-to-spot fossils. Another limestone of entirely different character can be found in the form of the hallway's circular radiator covers, which are made of the same "Black and Gold" Portoro seen in the entryway urns. Here, however, this black rock's stout veins, white to pale blue when filled with calcite and buff to yellowish when made of insoluble iron-rich clays, are much more vivid and accessible to close inspection.

16.8 The Brewster

2800 Pine Grove Avenue
Original name: Lincoln Park Palace
Completed in 1893
Architect: Enoch Hill Turnock
Geologic feature: Sioux Quartzite

Located just one block away from the previous site, this imposing bayed apartment building is the antithesis of the Elks National Memorial in both style and stonework. In contrast to the latter's bevy of elegant carved and polished stones, none of which is older than the Paleozoic era, the Brewster's exterior is monolithic, rugged, quarry-faced, and at least three times as ancient. The marine transgressions and tectonic upheavals that would create the Elks Memorial's fancy stone types would not occur until at least a billion years after the pinkish-purple rock here came into being. Geologists who have mapped and studied it have suggested its source sand was deposited in braided streams traversing an alluvial plain. This stream may have flowed through a basin created by recent thinning, rifting, and subsidence of the crust after the underside of ancestral North America had collided with an island arc and a separate microcontinent to create the Upper Midwest's Penokean Mountains.

Architects and suppliers in late nineteenth-century Chicago referred to this rock as "Jasper Stone," which derives either from the community of Jasper, Minnesota, or from the Jasper Stone Company, which ran quarries in that town and across the state border in Sioux Falls, South Dakota. Regardless, its geologic name is the Sioux Quartzite. Paleoproterozoic in age, the Sioux is usually a lighter pink or red, but some of its beds are indeed this purplish. To be petrologically precise, it's an orthoquartzite, which means that its extreme erosion- and weathering-resistant hardness is due not to significant metamorphism but to the fact its sand grains were from the beginning cemented with silica—in essence, quartz fused to more quartz. While quartzites can be too difficult to work to be good architectural stone, the Sioux is split easily along flat faces, and though it still can't be polished or easily carved, it makes an impressive rock-faced selection. Accordingly, it can be found adorning a number of homes and larger buildings in Minnesota and South Dakota. But to see it on a surviving building this far east is quite unusual.

16.9   Henry Rohkam House

1048 W. Oakdale Avenue
Completed in 1887
Architect: Theodore Karls
Geologic feature: Northwestern Terra Cotta

Built for the vice president of the Northwestern Terra Cotta Company, the Flemish Renaissance Rohkam House was unashamedly intended as a showplace for his firm's products. While the source of the cream brick that makes up most of the exterior has not been identified, the building's many terra-cotta details, including the wall in front, the west-side figurative panel, spandrels, string courses, window hoods, and trim definitely come from the Northwestern works. Its clay was dug not from a local source but from the Illinois Basin coalfields of both this state and Indiana. The Pennsylvanian-age shales and underclays found there were well worth shipping into the city, because they were free of glacially derived pebbles and performed well in the extra-high kiln temperatures needed to produce high-quality, fireproof terra-cotta.

16.10   Temple Sholom

3480 N. Lake Shore Drive
Completed in 1930
Architectural firms: Loebl Schlossman & Demuth, with Coolidge & Hodgdon
Geologic features: Lannon Dolostone, Salem Limestone

Though it's by no means as vertically imposing as many of the other buildings fronting North Lake Shore Drive, Temple Sholom is one of the most eye-catching architectural landmarks for motorists speeding by. It's an octagonal, Byzantine-Revival mass of stone mostly of a color a shade or two richer than the omnipresent Salem Limestone. But to fully appreciate both its geology and the wealth of its architectural detail, it's essential to exit the Outer Drive and take a much closer look.

There's no doubt that much of this building's external appeal lies in its designers' canny choice of complementing rock types. The Salem is present, but it's been relegated to a very effective supporting role as trim and the medium for the carved ornament. Most of the synagogue's exterior is clad instead in a Regional Silurian Dolostone. It suggests the texture and palette of our native Lemont-Joliet and Artesian rock, but in a way that's somewhat subtler than these. This is the Lannon, quarried in the town of that name that lies fifteen miles northwest of Milwaukee.

Unlike its geologic equivalents produced in northeastern Illinois, the Lannon Dolostone is still being produced, and it remains a popular choice for everything from architectural cladding and veneer to garden edging. After the demise of the

FIGURE 16.4.   Lake View's Temple Sholom is a handsome demonstration of the twentieth-century return of Regional Silurian Dolostone to Chicagoland architecture after the demise of local sources and the long reign of Indiana's drably dependable Salem Limestone. The West Side and Lower Des Plaines Valley producers were long gone, but the quarriers of Lannon, Wisconsin, eagerly provided similar stone. Now Lannon-clad mansions and houses of worship are a common sight throughout the region.

Chicagoland building-stone industry, early twentieth century architects, who'd perhaps had their fill of the dull anonymity of the Salem Limestone monoculture, turned to the Lannon, especially when it became apparent that it generally weathers more gently than its old Illinois counterparts. Its better resilience is sometimes purchased at the price of a more subdued play of colors. Still, the Lannon exudes a sort of luminous dignity, both cheerful and august, that makes it an almost irresistible choice for houses of worship, mansions, and humbler structures, too. This fact has not been lost on the architectural community. If you spend a day exploring Chicago's suburbs, you'll see just how much of a standard design formula the Lannon-Salem coupling has become in both religious and residential settings.

UPTOWN AND RAVENSWOOD

17.1 Graceland Cemetery

4001 N. Clark Street
Established in 1860
Geologic features: Graceland Spit, various rock and metal types
  described below

> The whole [of Graceland Cemetery] has a park-like appearance; and in
> this direction there is developed a beautiful effort to relieve to the utmost
> possible extent the extra solemnity which usually broods over places of
> sepulture. The glories of spring, and the rich fruition of summer are full
> of suggestions to the mourner who wanders through the shaded aisles,
> of a possible spring and summer which have no autumn nor winter.
> (Franc B. Wilkie, *Marquis' Hand-Book of Chicago*, 1885)

In the nineteenth century, before the introduction of backhoes and other
modes of powered excavation, the task of digging through our region's rock-
studded glacial till was not a pleasant one. For that reason, many Chicagoland cemeteries of that time were sited where the substrate is sandy and consists instead of
finer outwash or, if close to the lake, beach and dune deposits. While Graceland's
founders no doubt appreciated and took full advantage of both its high ground
and yielding soil, they could not have known that what would become the world-
famous resting place of many of Chicago's greatest architects and other notables

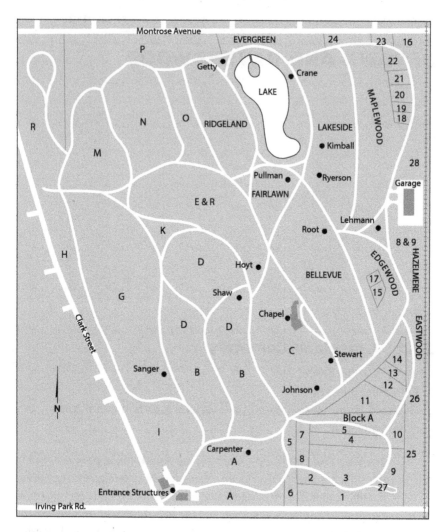

MAP 17.1. Graceland Cemetery.

is in fact situated on a landform of considerable geologic significance. Named the Graceland Spit after the cemetery where it is still extant, this broad if subtle ridge is a surviving segment of a gently curving deposit of sand and gravel that formed in the earlier Holocene between 5 and 4 ka, during Lake Michigan's Nipissing highstand phases. At its maximum the lake's surface was 20 feet higher than present. What are now the neighborhoods to the east and west of the cemetery were submerged in the waters of lake and lagoon, respectively. Originally the Graceland Spit extended in an unbroken arc all the way down to the South Side, but later, when the lake level dropped again, its central and most eastward portion was eroded away by surf action.

Still, the geology of this site hardly stops there. Inside its enclosing walls there's a multitude of different rock types, each with its own fascinating origin story. And the cemetery is also one great open-air laboratory of differential weathering, where more resistant varieties of stone have survived Chicago's harsh climate in fine form, while others initially attractive but ultimately less suitable have been ravaged by the elements, even in this protected, sylvan environment. The following tally of stops in the cemetery is by necessity just a sampling, but one that is based on both years of research and the various walking tours I've led here. For the exact location of each site, see map 17.1.

Entrance Structures (1896, Holabird & Roche)

This includes the gateway pylons, the Waiting Room to the left as you enter, and the Administration Building to the right. All are made of rock-faced Waupaca Granite, the one most striking selection of its type to be seen anywhere in the city. In the words of geologist Ernest Robertson Buckley, "There is probably no more beautiful or brilliant stone quarried anywhere in the country." He was writing of its use for interior decoration, but here his words are just as apt for exterior settings as well. This is a classic example of a porphyritic igneous rock that has *phenocrysts* (large crystals) set in a matrix of other minerals—in this case, quartz, biotite, and hornblende. The Waupaca's hefty phenocrysts are pink to brick-red alkali feldspars (microcline and orthoclase) that in places are mantled in a lighter-toned type of plagioclase. This coated-feldspar feature is called *rapakivi*, a term that originally referred to a similar type of granite found in Finland. It's not as apparent on rock-faced surfaces, but it is more visible at the Chapel, discussed below.

The Waupaca Granite is Mesoproterozoic in age and comes from central Wisconsin's Wolf River Batholith, a massive body of magma that formed from the partial melting of the lower continental crust 15 or 20 miles down. That magma rose almost all the way to the surface before solidifying. While the Waupaca is some 260 Ma younger than the Montello Granite, it shares the same classification as an alkali-feldspar granite, and the same enigmatic origin in what was otherwise a quiet continental interior, far from a subducting plate boundary or area of active mountain building. What caused the Wolf River Batholith to form may so far elude our understanding, but it's interesting that it is just one of a series of similar magmatic events that occurred at that time from the American Southwest all the way to eastern Canada.

Philo Carpenter Monument (date of installation and designer unknown)

This rather uncomplicated monument has a complicated record. The 1885 Rand-McNally guidebook states that it consists of "a plain column of Italian marble." The subsequent National Register of Historic Places description avers that it's a column of Lemont-Joliet Dolostone instead. It turns out that of these two sources the earlier one is the more correct, though the National Register entry isn't completely off base. Both the inscribed pedestal and the column above it have all the earmarks of Carrara Marble—it's white, fairly fine-grained, and faintly streaked with wisps of silvery gray. The sturdy base on which all this rests is another matter, however. It's badly weathered and encrusted, but in places its broken or spalled surfaces have turned that telltale buttery yellow, a characteristic giveaway that it's the Lemont-Joliet Dolostone.

The monument celebrates the man who was both a prominent abolitionist and Chicago's first pharmacist. It's also an indicator that this famous, Oligocene-age marble from Italy's Apuan Alps was widely available at a fairly early date in the city's history. Here the Carrara has done what can only be expected, given its prolonged exposure to the local climate and air pollution. It has weathered more than its makers intended, as is evidenced by its sugared texture, spalled sections, and degraded lettering of its inscriptions. Still, the degree of stone decay here is modest compared to some of the other older marble headstones in Graceland and other urban cemeteries.

Johnson Monument

This relatively modest site should not be confused with the more elaborate Jack Johnson burial plot in a different section. The pink and gray-swirled rock comprising the base and upright is the Paleoarchean (and, at 3.52 Ga old, staggeringly ancient) migmatite known as the Morton Gneiss. Quarried in southern Minnesota, the stone is well represented in Chicago architecture, and has also been a popular choice in the monumental trade. It can be spotted elsewhere in the cemetery as well.

John K. Stewart Mausoleum (ca. 1916, Hugh Price)

This tomb in the ancient Egyptian style is made from that most attractive pink form of true granite, Massachusetts' Neoproterozoic Milford. This is the same stone found on the Glessner House's exterior and other notable Chicago buildings. But this mausoleum and at least one other Graceland gravesite marker proves that it was also favored for use in the monumental trade. While currently it hosts a considerable community of lichen and green algae that clearly have good taste in the home they've chosen, the structure still has a few surfaces clean enough to reveal the Milford's black-biotite-speckled, medium-grained appearance. Also make sure that you take a look through the glass door. The interior is clad in the same stone, the Carrara Marble, seen above at the Carpenter Monument. But here, where it gets the protection from the elements it so dearly needs, it's of the "English-vein" variety.

The Chapel (1888, Holabird & Roche)

Nowadays the Waupaca Granite is so rare elsewhere in the city and is such a fascinating rock type in itself that it deserves a second look, on the exterior of

FIGURE 17.1.   The Graceland Chapel's distinctive Waupaca Granite, in both dressed and polished form (left and center) and rock-faced finish. Visible here is one classic characteristic of the type of granite known as rapakivi: in places the dark alkali-feldspar crystals are ringed with light plagioclase feldspar.

this lovely building, which the architectural historian Robert Bruegmann has described as "solid, almost primeval." Once again, the stone exterior is mostly in rock-faced finish, which at time of writing is considerably sootier in spots than it is on the entrance structures. Nevertheless, it's also found by the chapel entrance dressed-face and polished, where its rapakivi texture is on much better display. Here you'll easily recognize the white plagioclase jackets on the subrounded alkali-feldspar phenocrysts. The Wausau's two color variants, red-and-black and red-and-green, are both clearly visible. The first of these represents the original form of the granite; the latter is the more altered type, where some of the feldspar has changed to epidote, and the hornblende and biotite to chlorite.

John Wellborn Root Monument (1895, Charles B. Atwood & Jules Wegman)

Root, the illustrious designer of the Monadnock Building and the Rookery, died in 1891, at the tragically young age of 41. Four years later this Celtic cross was erected at his gravesite. That event was reported in both the *Inland Architect* and the *American Architect*; and each journal identified the stone used for this imposing monument as "Scotch granite." This strongly suggests that the rock type you see here is a selection of Aberdeenshire Granite, a product of Scotland's northeastern reaches that was in great demand in nineteenth-century America. When I checked with the British geologist and architectural-stone expert Gordon Walkden, of the University of Aberdeen, I learned that the cross could be a weather- or cleanser-bleached example of the pink to red "Peterhead" variety—which, as it so happens, also once adorned Root's ground-floor exterior of the Reliance Building. Or, alternatively, it might be a pink variety of the Aberdeenshire from the Deeside quarry district. The Aberdeenshire Granite's impressive array of varieties range in color from the Peterhead's distinctive ruddy tones to salmon, light pink, light gray, and bluish-gray. All these types date to the Devonian period, and were emplaced in the upper crust late in the mountain-building event known as the Caledonian Orogeny.

Ernst Johann Lehmann Mausoleum (1919, Mundie & Jensen)

While many of the mausolea in the cemetery are set in the neoclassical style, this one has the distinction of being clad in North Carolina's Mississippian-age Mount Airy Granodiorite. It's chastely white at a distance and salt-and-peppery up close. The Mount Airy is renowned for having been the replacement stone for the mammoth Aon Center's failing Carrara Marble, but it was in use long before that, and its use included monumental applications such as this. The surfaces here are in good shape, and you'll be able to pick out its white plagioclase and alkali feldspars, gray quartz, and black biotite.

George M. Pullman Monument (1897, Solon S. Beman)

This site is the grandiose resting place of the maker of the Pullman railroad sleeper car and creator of the Pullman community. As a 1900 *Inland Architect* writer noted, "the construction is of enduring and substantial character." That, it turns out, is an understatement. After the 1894 strike that rocked the Pullman works the man was such a detested figure in organized-labor circles that when he died just three years later his family thought it best to make his remains desecration-proof. As a result, Pullman's lead-lined casket, sheathed in tar paper and given an external 1-inch coat of asphalt, lies in a pit that is 8 feet deep, 13 feet long, and 7 feet wide. The pit's contents constitute an interesting stratigraphic succession. The bottom 18 inches is a floor of solid concrete upon which the casket rests. The next layer of concrete surrounds the casket and rises half an inch above its lid. At that level there's a set of transverse steel T-rails secured with two rods. On top of that, another layer of tar paper, and above that, the uppermost stratum of concrete. All this could make Pullman the one person in Chicago most likely to survive the forces of erosion and be collected as a fossil in some future geologic period.

Aboveground, the site is designed as a neoclassical exedra with one free-standing Corinthian column. The stone chosen was the fine-grained and aristocratically gray Hallowell Granite. Considerably more exposed than the person it honors, it nevertheless has held up well. This stone is derived from one of Maine's many Devonian batholiths that owe their origin to the formation of the Acadian Mountains, which in turn were occasioned by the collision of Laurentia (ancestral North America) with the microcontinent Avalonia. It's composed of the alkali feldspars orthoclase and microcline, the plagioclase-feldspar oligoclase, quartz, and the micas muscovite and biotite.

Martin Ryerson Tomb (1887, Louis Sullivan)

While the stone used here is also present on the Palmer House Hotel, this lumber-baron's sepulcher is the best place to see Massachusetts' remarkably dark-toned Quincy Granite. This striking selection was originally thought to be either Ordovician or Silurian in age, till a new and more tightly constrained age determination, made by measuring the relative abundance of uranium and lead isotopes in its zircon crystals, demonstrated that it's early Devonian instead. Granites, being felsic igneous rocks, are usually lighter in hue, but the Quincy's combination of bluish-gray microcline, darker-than-usual quartz, and black hornblende makes it a "black granite" that for once really is a granite, albeit of the alkali-feldspar type. Here it imparts a somber and timeless dignity that is the perfect choice for Sullivan's Egyptianate design.

William Kimball Monument (1904, McKim, Mead & White)

This neoclassical colonnade was made of the "Second Statuary" variety of Shelburne Marble, from West Rutland, Vermont. As is true of many older marble headstones situated closer to the cemetery entrance, this rock has been cruelly treated by acidic compounds in the rain and urban atmosphere, and now serves as a powerful lesson in the dangers of using this calcareous metamorphic stone in Chicago's outdoor settings. While some of the monument's marvelous ornamental detail still looks pristine, the face of the monument's presiding angel has partially dissolved, and there are also obvious signs of advanced chemical weathering on the low wall and column bases.

The Shelburne, once extracted in a number of localities up and down the Green Mountain State's western, Taconic flank, is a lower-Ordovician true marble still quarried today in Dorset Mountain, near Danby. It began as a carbonate-bank limestone on the continental shelf of Laurentia. Subsequently it and other sedimentary deposits were thrust far inland from their original position by an advancing arc of volcanic islands that collided with Laurentia to produce the Taconic mountain chain. In the midst of all this tectonic Sturm und Drang, the limestone was transformed into high-quality marble varying from pure white to heavily patterned with green and gray veins.

Richard Teller Crane Jr. Mausoleum (1927, John Russell Pope)

Designed by the architect of the Jefferson Memorial in Washington, DC, this elegant, Ionically becolumned tomb in its lovely site by Lake Willowmere gets my vote for the best neoclassical statement in Graceland. And its stone is the best selection possible: the Barre Granodiorite, quarried in the eponymous Vermont town. Dating to the Devonian period, it's one of several granitoids to be seen in Chicago that owe their origins to the Acadian mountain-building event. Very popular and extensively used in the monumental trade, the Barre is similar in mineral composition to the Lehmann Mausoleum's Mount Airy, but its greater biotite content makes it look even more peppery at close hand, and grayer at a distance.

Carrie Eliza Getty Tomb (1890, Louis Sullivan)

Usually considered Graceland's architectural crown jewel, this little wonder displays Sullivan's genius for ornamental detail in green-patina bronze and that

FIGURE 17.2.   American cemeteries are, among other things, the final resting places of disjointed neoclassical references. But once in a long while, as in this leafy lakeside setting at Graceland, the old Greco-Roman magic works its spell again. Part of the Crane Mausoleum's valid claim to perfection lies in the stone its architect selected—the soothingly sober, salt-and-pepper Barre Granodiorite.

mason's delight, the Salem Limestone. Unlike the Shelburne Marble cited just above, this southern Indiana staple of the building trade, often called "Bedford Stone," is an example of a limestone that formed deep in the stable continental interior and so was never subjected to the crustal stresses that metamorphose this rock type into marble. But while it too is composed largely of calcite, the Mississippian-subperiod Salem resists weathering to such a degree that it often looks freshly cut after many decades of exposure to the rain and wind.

William M. Hoyt Monument (1883, designer unknown)

With its three allegorical women commanding the ramparts, this ostentatious Victorian monument is hard to miss along the roadway. While the cemetery's National Register of Historic Places listing states that it dates to about 1904, supplier's order-book records in the archives of Westerly, Rhode Island's Babcock-Smith House Museum indicate it was made and shipped two decades before that. And its gray stone, known to geologists as the Narragansett Pier Granite, did come from that fine town. In fact, the Narragansett Pier, which must have been extremely popular in its day, can also be found in a large number of other Graceland mausolea and monuments. Fine-grained and apparently very enduring, the "Blue Westerly" (as it was marketed) is frankly not the most striking granite to be found on the grounds, but it's a uniform and dependable whitish-gray. The Narraganset Pier solidified from magma in the Permian, as part of the formation of the great Alleghenian mountain chain, as Gondwana merged with what is now eastern North America during the assembly of Pangaea. If you inspect the stone with a hand lens you'll spot its light-toned microcline and oligoclase feldspars, gray quartz, and black biotite.

Howard Van Doren Shaw Monument (1926, Howard Van Doren Shaw)

One of the most notable Chicago architects of his day, Shaw designed this column surmounted by a bronze sphere. The latter has weathered nicely to produce the customary green, copper-salts coating. The column is a splendid polished example of the "Redstone Green" brand of New Hampshire's Conway Granite. The faint olive tint is easier to recognize at a little distance, but up close the details of this beautiful rock are also worthy of scrutiny. Crystals of gray orthoclase, the alkali feldspar, are intergrown with white plagioclase feldspar. The quartz, which gives the rock its unusual tint, is olive-tan. Black biotite and hornblende are also present. Jurassic in age and a product of the breakup of the supercontinent Pangaea, the Conway is a mere youngster compared to the other igneous rocks used for Graceland monuments.

William McKibben Sanger Monument (ca. 1905, George R. Dean & Arthur R. Dean)

When the rhyolite block chosen for this monument was lifted out of its Berlin, Wisconsin, quarry, it weighed between 15 and 20 tons, according to a local newspaper. It was then shipped to Chicago for finishing, which included a high-grade polish. Now adorned with a bronze rendering of a praying female figure set on a Celtic cross, this great pyramidal slab certainly appears to be a rock for the ages. To once again quote Ernest Robertson Buckley: "The Berlin rhyolite is the strongest, and one of the most durable building stones on the market. The polished surface of a column will not be perceptibly injured after centuries of exposure." For that reason, it was a favorite selection for cemetery monuments, and was also used widely as long-lasting road pavers.

The Berlin Rhyolite is the product of a massive volcanic event that occurred about 1.76 Ga, at the same time the mineralogically similar Montello Granite was emplaced some 20 miles to the southwest. On the back of the monument there is a strong suggestion of the flow-banding texture that is often found in this rock type. The rhyolite's small pink feldspar phenocrysts are also evident.

17.2   Aragon Ballroom

1106 W. Lawrence Avenue
Completed in 1926
Architectural firm: Huszagh & Hill
Geologic feature: American Terra Cotta

This historic performance venue is clothed in an exterior described as Moorish, and it's well worth a stop to behold its highly ornate terra-cotta designs executed by the American Terra Cotta Company based in McHenry County, Illinois. The building makes a good point of comparison with the similarly flamboyant Garfield Park Fieldhouse, which also features beautifully crafted ornamental cladding produced by the superbly skilled artisans of the same firm.

17.3   Old Town School of Folk Music

4544 N. Lincoln Avenue
Completed in 1931
Original name: Frederick H. Hild Branch of the Chicago Public Library
Architect: Pierre Blouke
Geologic feature: Morton Gneiss

Most of this building is clad in undistinguished brown brick of unknown source, but its entrance is flanked by panels of Morton Gneiss, a great favorite of Chicago's Art Deco designers. With an isotopic age of 3.52 Ga, this highly contorted, Paleoarchean migmatite is by far the oldest used in American architecture. It can also be seen just a mile eastward at Graceland Cemetery, where it assumes its other architectural role as monumental stone.

17.4   Krause Music Store

4611 N. Lincoln Avenue
Completed in 1922
Current name: Studio V Design
Architects: William Presto with Louis Sullivan
Geologic feature: American Terra Cotta

As Carl Condit wrote in *The Chicago School of Architecture*, "When we see a building of Sullivan's, we see not only structure and material form but also the creator's own inner world of emotions and dreams. We are compelled to share them." The Krause Store is the North Side's most awe-inspiring display of both Sullivan's inner world and American Terra Cotta Company craftsmanship.

The great architect's contribution, his last realized work, was this building's façade, which he designed in an office space the American Terra Cotta management provided in his grim final years. To me, the botanical flight of fancy on view here is the best place anywhere to intimately appreciate the master's astounding organic geometry. It seems to rise out of its surface as though the molecules of clay and glaze are arranging themselves in a spontaneous evolutionary process.

EDGEWATER, ROGERS PARK, AND SAUGANASH

18.1   Epworth United Methodist Church

5253 N. Kenmore Avenue
Completed in 1891
Original name: Epworth Methodist Episcopal Church
Architects: Frederick B. Townsend (1891), Thielbar & Fugard (1930)
Geologic features: Fieldstone, Lemont-Joliet Dolostone, Concrete

Houses with exteriors of Fieldstone—that is, with boulders collected from farm fields that are mounted, usually random-coursed, in a matrix of mortar—are quite a common sight in the glaciated Upper Midwest. During the flowering of the Richardsonian Romanesque, Fieldstone was also used, often to stunning effect, for churches, as can be seen here and in such communities as Batavia, Illinois, and Lake Geneva, Wisconsin. But on the eastern side of Chicago, where the substrate is composed of boulderless beach and ancient lake-bottom sediments, there is no ready supply of suitably large stones. So in the case of this striking church the deficiency was remedied by shipping the requisite rocks, erratics harvested in Wisconsin, using what was then the most convenient mode of coastal transportation for bulk cargo: Lake Michigan itself. The stones were delivered to a special slip constructed at what was then the foot of Berwyn Avenue, just a short distance from the worksite.

Erratics are fragments of rock that have been torn from the Earth's crust or from surface deposits by the plucking action of a passing glacier. In the Pleistocene ice age, huge quantities of rock were transported southward from such locales as Ontario, Michigan's Upper Peninsula, and northern Wisconsin. But until the Swiss-born Louis Agassiz and others convincingly demonstrated that ice sheets had covered this and other parts of the northern United States, early geologists, using the Bible as their primary historical-geology text, assumed that the erratics were dropstones melted out of icebergs adrift in Noah's flood.

The stones present here represent at least several varieties of rock that came from distant stations northward. The one most prevalent boulder type is basalt, which very likely came from the Mesoproterozoic Midcontinent Rift lava flows of the Keweenaw Peninsula. Also visible are red and gray granites, chunks of Silurian dolostone, and a wavy-textured form of metamorphic rock called schist. Quarried Lemont-Joliet Dolostone, also Silurian, was used for the sills, trim, and stringcourses of the original church section. When a new front entrance was constructed at the same time as the major 1930 addition, that rock type was substituted by color-coordinated, pigmented concrete.

18.2   Rosehill Cemetery

5800 N. Ravenswood Avenue
Established in 1859
Geologic features: Rosehill Spit, Lemont-Joliet Dolostone, Hallowell
    Granite, Narragansett Pier Granite, Morton Gneiss, Carrara Marble,
    Colorado Yule Marble, Grueby Terra Cotta Faience

As mentioned in the description of Graceland Cemetery (site 17.1), early Chicagoland graveyards were often selected for their sandy, easily excavated soil. High and dry terrain was also much preferred. For these reasons, it's not surprising that the city's two most famous landscaped burial grounds are found situated astride spits—long, arcuate deposits of sand and gravel deposited by coastal currents during highstands of earlier versions of Lake Michigan. But while the Graceland Spit formed in the relatively recent Nipissing phases, the Rosehill Spit is over twice as old. It came into being about 11.5 ka ago, during the earliest-Holocene Calumet stage, when the lake's surface was some 40 feet higher than now. Consequently, at a maximum altitude of a little over 620 feet above mean sea level, Rosehill Cemetery sits on average about 20 feet higher than Graceland. However, even its ridge is not the most elevated in the greater metropolitan area. The Wilmette Spit, located just to the north, formed at 640 feet during the even earlier Glenwood II stage 12.2 ka ago, just over the line into the Pleistocene. Be that as it may, early area inhabitants recognized this site as distinctly high ground and called it Roe's Hill. Only later did a bureaucratic error on an official document render it "Rose Hill," which later further elided into "Rosehill." It's a

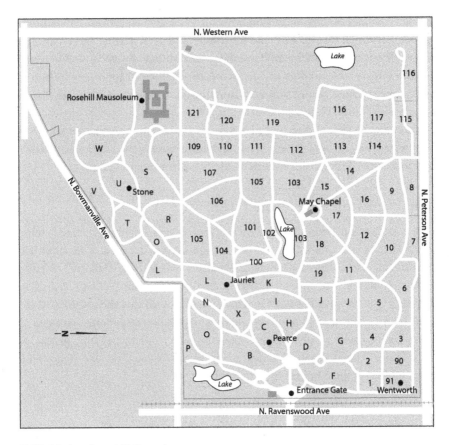

MAP 18.1. Rosehill Cemetery

splendid example of a beneficial mistake. This more poetic and flowery version just sounds more appropriate for a cemetery.

I've selected the following points of interest at Rosehill Cemetery partly on the basis of their well-documented stone types, but also for their overall geologic significance. Their locations are indicated on map 18.1. You'll also find that the other monuments and headstones here feature beautiful if not as well-provenanced examples of igneous, metamorphic, and sedimentary rock types, and that a stroll or drive through the whole complex is both geologically and historically rewarding.

Entrance Gate (1864, William W. Boyington)

If you're familiar with the Chicago Water Tower and Pumping Station, the cemetery's monumental portal is likely to give you a pronounced case of déjà vu. There is the same buff-and-yellow-tinted Lemont-Joliet Dolostone, and the same architect's unquenchable obsession with turrets and castellation. This Silurian sedimentary rock formed in shallow seawater and on a narrow carbonate shelf set on the edge of the Michigan Basin about 425 Ma ago. By the mid–nineteenth century our region's builders and architects had learned to appreciate its beauty, utility, low cost, and ease of shipment from the quarries of the Lower Des Plaines River Valley via the Illinois and Michigan Canal. And in the days before the devastating fires of 1871 and 1874, and before the expansion of the railroad network neutralized the Lemont-Joliet's local advantage, it was the city's predominant building stone. Today, it survives in all too few of our major buildings.

John Wentworth Monument (1888)

First a Jacksonian Democrat and then a Lincoln-era Republican, "Long John" Wentworth is one of a full dozen of Chicago mayors interred at Rosehill. His monument, a tall obelisk, is made of Maine's Devonian-age Hallowell Granite, a very popular architectural and monumental stone in late 1800s Chicago. The February 1889 issue of the trade journal *Stone* reported that the obelisk's 70-ton shaft took 2 hours and 40 minutes to raise and place on its pedestal, and that all told the monument weighs 180 tons and stands 66 feet 6 inches from ground to apex. (Later accounts make it 72 feet tall instead.) As you'll see, the Hallowell is gray and fine-grained but subtly porphyritic.

Frances Pearce Monument (sculpted in Rome, Italy, in 1856 by Chauncey B. Ives, first located in a cemetery in what is now Lincoln Park, moved to Rosehill Cemetery in 1861)

Among other things, this grand expression of Victorian sentiment is a sterling lesson in differential weathering. The reclining Carrara Marble figures of mother and daughter, apparently encased in their glass viewing box since the monument's transfer to Rosehill, retain their sharp detail and luster, but the pedestal beneath, afforded no such protection, is badly worn and sugared, with some of its inscriptions now considerably less legible. Compare them with the still distinct sculptor's name incised into the slab end beneath Frances's bepillowed head. Also note how the copper-salts weathering product of the metal box frame has stained it in places a light green.

Charles Fabian Jauriet Monument (1886, designer unknown)

Of all the Rosehill gravesites that feature the "Blue Westerly" variety of Rhode Island's Narragansett Pier Granite—and there are at least 17 of them—this

FIGURE 18.1.   Rosehill's Pearce Monument is one indicator of the early use of imported Italian Carrara Marble in Chicago. But always vulnerable to acidic rain and pollutants and to this climate generally, true marbles often do not fare well when fully exposed. Here the unprotected base has suffered considerable deterioration while the encased sculpture has not.

monument, topped with a graceful female figure gesturing upward, is the most immediately appealing, and also the easiest to locate. The Narragansett Pier, Permian in age, is a very fine-grained off-white granite that made up for its relative lack of visual interest by its dependability; I've yet to see one dating from the 1880s that isn't still in tiptop shape. It was certainly one of the leading monumental stone selections of its time.

Horatio L. May Chapel (1899, J. L. Silsbee)

This handsome structure, solidly built of an undocumented gray granite, features a terra-cotta treasure within, in the form of decorative faience created by the Grueby Company of Revere, Massachusetts. Chicago boasts no finer example of how common clay, when burnt and enameled, can provide tremendous ornamental impact. Note especially the stunning ceiling design of the porte cochère (covered entryway).

    Incidentally, the cemetery's ponds, including the one that lies adjacent to the May Chapel's entrance, have a geologic significance all their own. According to

FIGURE 18.2.   Gazing upward at the remarkable Grueby faience mosaic in the May Chapel's porte cochère.

the 1885 *Marquis' Hand-Book of Chicago*, they are, or at least originally were, fed by an artesian well that, bottoming at 2,279 feet, was dug almost half a mile down into the Earth's crust. In all likelihood, the aquifer tapped was the Cambrian Mt. Simon Sandstone, which sits just above the ancient basement of Proterozoic igneous rock.

Norman H. Stone Mausoleum (year and designer unknown)

This more modern site is a small structure, simple in its lines but striking in its use of Morton Gneiss, the Paleoarchean migmatite that has long done double duty as both an architectural and a monumental stone. By far the oldest rock type on display in the city, it dates to 3.52 Ga. Here it is used not only in the wildly patterned exterior walls but in the urns flanking the entrance and the adjoining bench thoughtfully included for visitors. Note especially the mausoleum's back and sides, where waves of tonalitic gneiss and amphibolite clasts undulate across the surface as evidence of the extreme metamorphism this rock has undergone over the eons.

Rosehill Mausoleum (1914, Sidney Lovell)

Here the main geologic attractions lie indoors, in the form of two gleaming white marbles—one from the Old World, one from the New. The first, the renowned Italian Carrara, is present in the flooring and apparently in the columns and walls of the rotunda and entryway. It is mostly veinless and resembles this classic stone's finest, Statuario grade, as does the Pearce Monument cited above. But in this interior location, where it's protected from the elements, the Carrara is accorded its more appropriate role. A Jurassic limestone that was transformed into marble in the Oligocene, it is the product of large-scale regional metamorphism caused by crustal compression and mountain building over a very large area. This most prestigious of marbles has a history of human use that stretches back to the reign of Augustus, the first Roman emperor, when its quarrying locality, dedicated to the goddess of the Moon, was known as Luna, and its stone, *marmor lunense.*

The Carrara is graciously complemented by a handsome gray- and golden-veined variety of Colorado's Yule Marble, which makes up much of the other wall cladding. Mississippian in age, it is in contrast the product of contact metamorphism. This much more localized process, a less frequent source of commercial marble, is the result of the original stone being cooked by a nearby intrusion of a body of magma. Elsewhere the Yule's parent rock, the bluish Leadville Limestone, has remained unchanged. Since the beginning of the twentieth century the Yule has been extracted from five different open and shaft-mine quarries above Yule Creek, in the heart of the Rockies. The only human settlement in the area is predictably named Marble.

A wholly unbiased person evaluating these two beautiful stone types would find it exceedingly difficult to say that one or the other is the classier. This fact is not lost on Carrara's marble experts themselves, some of whom are now running the one Yule quarry currently in operation. Remarkably, 90 percent of the stone they're extracting is exported to Italy—which might seem to be the quarryman's equivalent of carrying coals to Newcastle. But these producers and marketers, whose families have been in this business for centuries, definitely know what they're doing.

18.3   Piper Hall, Loyola University

970 W. Sheridan Road
Completed in 1909
Original name: Albert G. Wheeler House
Architect: William Carbys Zimmerman
Geologic feature: Shelburne Marble

Examples of true marble faring badly outdoors are all too common in Chicago, as Graceland Cemetery's Kimball Monument and the Aon Center attest. However, the rock-faced exterior of this comely former residence, sited in the most exposed lakefront location conceivable, seems to have fared well. It's made of the bone-white "Vermont Danby" variety, derived from the same Shelburne Formation from which the Kimball Monument's West Rutland type was extracted. In the case of Piper Hall, the Shelburne Marble was quarried within Dorset Mountain in the Taconic Range on the Green Mountain State's western fringe. Lower Ordovician in age, it began as deposits of lime mud on a carbonate shelf off the coast of Laurentia, North America's Paleozoic equivalent. These deposits eventually became solid limestone, which during the Taconic Orogeny were thrust onto the continent by a colliding volcanic-island arc and metamorphosed into this beautiful and premium-quality marble, as good as any Europe has to offer. It's still quarried today by the Vermont Marble Company. That firm's Vermont Marble Museum, located in the town of Proctor, is open to the public and well worth a visit.

18.4   Mundelein Center, Loyola University

6363 N. Sheridan Road
Completed in 1931
Architects: Nairne W. Fisher and Joseph W. McCarthy
Geologic features: Salem Limestone, Sylacauga Marble, Holston
    Limestone, Botticino Limestone, Buixcarró Limestone

Resembling a Loop 1930s skyscraper that drifted like an iceberg up to Rogers Park and was then miniaturized, the Mundelein Center has almost all the right credentials for the Grand Art Deco Formula. The setbacks are certainly there, as is the obligatory Salem ("Bedford," "Indiana") Limestone exterior. The only thing missing is a thoroughgoing base course of less porous crystalline rock; and even here the outer wall of a more recently installed handicapped ramp is an unidentified gray granite.

If you walk up that ramp and scrutinize the Salem, you'll see that in patches the stone has deteriorated; it's spalling and showing salt efflorescence in places. However, the fact that this Mississippian-age biocalcarenite is accessible near ground level also gives you the chance to see the wonders of ancient life embedded within it. An unsigned advertising-section article in an 1897 issue of the *Inland Architect* noted that this rock type is, "when quarried, rather soft, but the air acts on this stone like water in tempering steel; the moment the air strikes it, it hardens but retains a certain amount of elasticity and when struck by a hammer resounds like the tone of a bell. . . . Upon close examination with a magnifying glass, the grain of this stone proves to be infinitesimal shells and fragments all bound together by a firm and even setting of lime carbonate. No art of man could construct a mass at once so firm, even and workable and at the same time so elastic and strong." Can any modern ad agency produce promotional prose this lucid and tintinnabulously poetic?

The Salem Limestone, quarried in southern Indiana, has also long been renowned for its ability to take and tenaciously retain the intricacies of detailed sculpting. Here the main entrance is dominated by massive renderings of the archangels Jophiel and Uriel, designed by Charles Fisher and realized by the artisans of the North Shore Stone Company.

In contrast to the monolithic exterior, the Mundelein Center's ground-floor interior has an impressive assortment of well-documented carbonate rock types. The checkerboard-pattern floors feature alternating pavers of Alabama's white Sylacauga Marble, of early Paleozoic age, and the widely used pinkish-brown Ordovician Holston Limestone, known in the building trades as "Tennessee Marble." The baseboards are the deep-green and white-veined Tinos Ophicalcite, a product of a Greek Aegean island and of Cretaceous age. The walls of the main hallway are clad in Italy's Jurassic Botticino Limestone; those of the vestibule just in from the main entrance feature the less frequently encountered "Florida Rose Marble" variety of Buixcarró Limestone. This rosy-beige and gray-veined Cretaceous sedimentary rock is quarried in the hills some thirty miles south of Valencia, Spain.

18.5   Queen of All Saints Basilica

6280 N. Sauganash Avenue
Completed in 1960
Architects: Meyer & Cook
Geologic features: Lannon Dolostone, Salem Limestone, Crossville
    Sandstone, Rosso Ammonitico Veronese Limestone, Levanto
    Ophicalcite, unidentified Porphyry

While it bears a close resemblance to Chicago's early churches made of our region's own Lemont-Joliet Dolostone, this imposing basilica is a product of the Space Age. Its style might best be described as Retro Gothic. The exterior here directly hearkens back to River North's Holy Name and St. James Cathedrals, but it's actually clad instead in Lannon Dolostone, a latter-day stand-in for our city's original and now sadly unavailable building stone.

The Lannon is present here in its most characteristic form, as random-coursed ashlar, and one can see why twentieth- and twenty-first-century architects, weary of the ubiquity and totalitarian blandness of the Salem Limestone, have often been eager to return to the city's Silurian-dolostone roots. The Lannon, quarried in a town on the outskirts of Milwaukee, is Wisconsin's widely used equivalent to the rock once quarried in the Lower Des Plaines Valley and on Chicago's West Side. If the compact-textured and dependable Lannon mostly lacks the weathering vulnerabilities of the original Lemont-Joliet stone, it also tends to be more muted in its coloration, with its bone-white to bluish-gray base color yielding less dramatically over the years to cream and buff and yellow. Still, it offers a warmer and somehow more locally appropriate alternative to Indiana's Salem ("Bedford") Limestone, which here serves very suitably instead as trim. This Lannon-Salem combo can be found again and again in our region's homes and houses of worship, where it has long been a dominant motif of building-stone use.

The basilica's roofing tile, documented as quarried in Vermont, is also of interest. Its tint and weathering characteristics suggest it's the Vermont Sea Green Slate type. Like the other color variants produced in the Green Mountain State, this stone originated as mud deposited in relatively deep ocean water in Neoproterozoic or Cambrian time that in the Ordovician was thrust up onto the margin of Laurentia and metamorphosed.

If it's open for visiting, also enter the church by way of the porte cochère, where you'll find that the adjoining spaces of the vestibule, narthex, and baptistry contain an spectacular array of ornamental stones. Those of the vestibule are not documented, apparently, but once you continue into the narthex you'll note the striking flooring and baseboards, which include an unidentified green serpentinite or ophicalcite and two different versions of the "RAV"— the nodular Rosso Ammonitico Veronese Limestone—in both its familiar red form as well as the wonderfully complementing yellow "Nembro Giallo di S. Ambrogio" variety. This Jurassic carbonate rock type from northern Italy is also found in the Loop on the exterior of the Pittsfield Building in a more obviously fossiliferous form. And the narthex walls feature the distinctly contrasting "Crab Orchard Stone"—the early Pennsylvanian-age Crossville Sandstone from the Cumberland Plateau of eastern Tennessee. This thin-bedded, fine-grained, varicolored, and interestingly patterned clastic sedimentary rock is still actively quarried. Its impressive hardness, visual appeal, and availability have long made it a popular choice for veneer and flagging, both out of doors and in. The Crossville is composed of grains that are so well cemented with silica that in decades past it has sometimes been described as a quartzitic sandstone or a quartzite. Using modern terminology, this would make it an orthoquartzite, the type that has not undergone any significant metamorphosis since it formed. On these walls you'll see the Crossville is marked by Liesegang rings, dark-brown lines that mark where iron-oxide minerals transported by percolating groundwater precipitated out of solution.

Just beyond lies the baptistry. Among its various beautiful rock types the best identified is the raised platform on which the font stands. It's made of the most unusual example of ophicalcite, the purple and white-veined Levanto, from Italy's La Spezia region. Within it float clasts of darkest green and black. Probably Jurassic in age, this rock does not conform to the usual greener tones of its equivalents. At some point in its long career underground it came into contact with hydrothermal fluids that transformed its coloration by carbonating the serpentine and oxidizing the magnetite into hematite.

The font itself is a red porphyry—a handsome igneous extrusive rock peppered with lighter feldspar phenocrysts. I presume it's not the famous "Porfido Rosso" quarried in Egypt by the ancient Romans and later widely reused by European stonemasons of the Middle Ages and later. That's usually purple or maroon. The brighter-toned stone here could have been quarried much more recently in a number of different locations.

Glossary

ABBREVIATIONS

cm = centimeters

Ga = billion years, or billion years ago

in = inches

ka = thousand years, or thousand years ago

km = kilometers

Ma = million years, or million years ago

mi = miles

mm = millimeters

Underlined items in the text have their own glossary entries.

Acadian Orogeny The orogeny that occurred in the Devonian period when Avalonia and other crustal fragments collided with Laurentia.

Acroterium (also rendered Acroterion) A vertically projecting ornamental element placed atop or on the edges of a pediment. The plural form of both singular spellings is acroteria.

Aggregate Particulate material such as sand, gravel, or rock fragments that is mixed with cement to produce concrete.

Alkali Feldspars One of the two main groups of feldspar minerals. Its members are the potassium feldspars (orthoclase, microcline, and sanidine) and anorthoclase.

Alkali-Feldspar Granite A granitoid rock that differs from regular granite in that at least 90 percent of its feldspar content is in the form of alkali feldspars.

Allochthonous Referring to a terrane that has been moved from its place of origin to a new location.

Amazonia An ancient continent that now forms a significant portion of South America.

Ammonite A type of cephalopod especially prevalent in the Mesozoic era.

Amphibole Referring to a group of silicate minerals of various colors found in some igneous and metamorphic rocks.

Amphibolite A green to black, weakly foliated metamorphic rock that has a variety of parent rocks, including basalt and komatiite.

Anorthoclase A silicate mineral of the alkali-feldspar group. It sometimes exhibits the property of labradorescence.

Anorthosite A felsic igneous rock that is composed almost completely of plagioclase feldspar, but may also contain in some cases significant amount of pyroxene. It often exhibits labradorescence.

Anthropogenic Produced by human activity; manmade.

Aphanitic Composed a microscopic crystals.

Aplite A granite of unusually fine-grained texture.

Apse The end section of a church interior, often semicircular in plan, that houses the altar.

Archean Eon The span of geologic time from 4.0 to 2.5 Ga.

Art Deco A term, derived from the Exposition des Arts Décoratifs et Industriels Modernes, for the architectural and decorative style prevalent especially in the 1920s. Its salient features include an emphasis on elegant, stylized ornamentation and streamlined and rectilinear forms inspired by machines and technology rather than by natural or organic forms.

Art Moderne The term referring to the leaner, less highly ornamented 1930s offshoot of Art Deco.

Ash Tephra that has a particle diameter of 2 mm or less.

Ash-Flow Tuff Tuff formed from ash laid down by a pyroclastic flow.

Ashlar Quarried stone that has been "squared off" (cut with sides that meet at right angles). Ashlar blocks may be arranged in regular courses or in a more random pattern.

Augite A dark-colored mineral of the pyroxene group.

Autochthonous Referring to a terrane that has remained in its original setting.

Avalonia The microcontinent that apparently formed on the margin of Gondwana but later broke off and migrated to collide with both Baltica and Laurentia, where it triggered the Acadian Orogeny.

Axed Finish A finish applied to stone in which the surface is rendered fairly flat but still rough to the touch using an axe, hammered or power chisel, or similar tool.

Baltica The ancestral version of northern Europe that existed in Proterozoic and Paleozoic time.

Baluster A small and often decorative column or pillar that supports either a horizontal or staircase handrail.

Balustrade A series of balusters surmounted by a handrail.

Banded Iron Formation A chemically precipitated sedimentary rock, most usually dating to the Archean or Paleoproterozoic, composed of alternating bands of chert and hematite or other iron-containing minerals.

Barrel Vault A ceiling of half-cylindrical form.

Basalt An aphanitic, quartz-poor, and usually dark-colored igneous rock. It is the extrusive equivalent of gabbro.

Bascule A counterweighted architectural element, such as a bridge span, that rotates around a horizontal hinge line.

Base Course The lowest course of cladding or masonry on a wall.

Bas-Relief A figure or image that projects in shallow relief from a wall or other surface.

Batholith A large pluton at least 40 square mi (100 square km) in area at the surface.

Bauxite A sedimentary rock that often forms from tropical soils or the weathering of carbonate strata. Of great economic significance, it is the primary ore of aluminum.

Bay An architectural element that projects outward from the main wall.

Beaux Arts Style An architectural style incorporating such classical elements as columns, pilasters, arches, domes, balustrades, medallions, and other Greek- and Roman-derived ornament to produce a sense of grandeur, symmetry, and formalism.

Bedded Layered.

Bedding A synonym of strata.

Bedrock Any section of rock that is still attached to the crust.

Biocalcarenite A type of limestone chiefly composed of very small shell fragments and tiny whole fossils.

Biotite A black or dark brown, sheet-forming silicate mineral of the mica family. One of its listed chemical formulas is $K(Mg, Fe)_3(AlSi_3O_{10})$; other variations are cited.

Bioturbation The disturbance or reworking of sediments by organisms.

Bisque In the terra-cotta trade, the fired clay that serves as the base for the glaze, which is subsequently fixed with a second firing.

Bitumen A black, tarlike substance made of hydrocarbons and derived from organic remains.

Black Granite A term, most usually an oxymoron, used widely in the quarrying and architectural trades. True granite by its mineralogical composition cannot be thoroughly black, though on rare occasions it can be rather dark-toned. This misnomer is used for such igneous rock types as gabbro, diabase, basalt, and some forms of anorthosite.

Book Match The symmetrical pattern created when two panels of cladding with identical patterns (usually cut in the quarry from the same block of stone) are mounted on a wall side by side, with one panel facing in the opposite direction to the other's. This creates an effect of an opened book with two mirror-image pages visible.

Boulder A detached rock larger than 256 mm in diameter.

Bowen's Reaction Series A description formulated by geologist Norman Bowen that shows the sequence in which silicate minerals crystallize as a magma cools.

Breccia A clastic rock composed of coarse, angular rock fragments.

Brecciated Referring to a marble or other stone type that features large angular or pointed rock fragments separated by an extensive network of veins.

Brick A shaped, rectangular unit of building material made of fired clay (which is sometimes combined with other substances).

Brownstone An architectural term for a red-, brown-, or maroon-tinted sandstone. This type of dimension stone was particularly popular with architects who adhered to the late nineteenth-century Richardsonian Romanesque style.

Brushed Finish A finish created by treating a stone surface with plastic or metal brushes to remove softer portions of the stone. This often imparts an aged look to the stone.

Bush-Hammered Finish A finish created by treating a stone surface with a multipointed hammering tool. This produces a rugged, nonskid surface.

Caisson As used in Chicago-region architecture, a vertical shaft, well, or pier dug underground and filled with concrete or other hard material. This shaft serves as an anchor which connects a building to the underlying hardpan or to the bedrock beneath it.

Calcareous Containing calcium carbonate.

Calciphilic Referring to organisms that thrive in calcium-rich or limey environments.

Calcite The carbonate mineral composed of calcium carbonate ($CaCO_3$).

Caldera A broad, bowl-shaped landform created by the collapse of a volcano due to the emptying of its underground magma chamber.

Caledonian Orogeny The early Paleozoic orogeny that resulted from the collision of Laurentia and Baltica.

Cambrian Period The span of geologic time from 541 to 485 Ma.

Capital The uppermost section of a column.

Carbonate Mineral A member of a chemical group of minerals that contain, among other substances, CO_3.

Carbonate Platform A marine shelf mantled in carbonate sediments. Carbonate platforms exist only in tropical or subtropical waters; one modern example is the Bermuda Banks.

Carbonate Rock A rock primarily composed of such carbonate minerals as calcite (calcium carbonate) or dolomite (calcium-magnesium carbonate).

Carboniferous Period The span of geologic time from 359 to 299 Ma.

Cartouche A decorative design, often set on a panel, tablet, or cladding unit.

Case Hardening The weathering process by which a rock's outer portion becomes harder and shelllike, while its interior becomes softer or crumbly. This is often due to the outward migration of mineral-laden water previously absorbed by the rock.

Cement A liquid substance used in the production of concrete that, when mixed with aggregate or other materials, sets and binds them together. Modern cement is usually composed of lime, gypsum, and clay.

Cephalopod A marine mollusc of the class Cephalopoda. Members of this class include the nautiloids, octopi, squid, and the now-extinct ammonites.

Chalcocite A sulfide mineral with the formula Cu_2S. It is a major source of commercial copper.

Chalcopyrite A sulfide mineral with the formula $CuFeS_2$. It is a major source of commercial copper.

Chamfer The beveled or recessed edge of a block of stone.

Chemically Precipitated Rock Referring to a sedimentary rock type that is formed by the precipitation of mineral crystals from water.

Chert A very hard, chemically precipitated sedimentary rock composed of microcrystalline quartz.

Chicago Window A classic component of Chicago office-building design, this is a rectangular horizontal element composed of one broad, unopenable center window panel flanked by two narrow, openable sash windows.

Chlorite A term referring to a group of often greenish silicate minerals associated with low-grade metamorphic rocks.

Churrigueresque Referring to a flamboyant version of the Spanish Baroque architectural style developed by the architects of the Churriguera family.

Cipollino An Italian term that literally means "resembling chives or baby onions." It refers to marble that has a banded or thinly layered appearance.

Cladding Stone used on the exterior of building for decorative effect. Usually it is not a load-bearing element.

Clast A rock particle or rock fragment.

Clastic Rock Referring to a sedimentary rock type that is formed of cemented particles—boulders, cobbles, pebbles, sand, silt, or clay—originally produced by the weathering or erosion of other, older rock.

Clay A mineral particle less than .003 mm (.0001 in) in diameter.

Clinopyroxene A subcategory of the pyroxene minerals that includes diopside and augite.

Coarse-Grained Referring to igneous or metamorphic stone types whose feldspar crystals are greater than 1.0 cm in length.

Cobble A rock particle 64–256 mm (2.5–10 in) in diameter.

Column A pillar, usually circular in cross section, that serves as either a load-bearing or merely ornamental element.

Common Brick Brick, usually of lesser ornamental appeal and made of unscreened or less carefully selected clay, that is used for building sides and rears, and for interior walls not intended for display. It is often softer, less sharp-edged, and less regular in shape than face brick.

Composite Order The classical architectural order that has capitals combining the volutes of the Ionic order with the acanthus leaves of the Corinthian order.

Conchoidal Referring to a mineral or rock that fractures to form curved or scooped-out surfaces. Quartz is a very common mineral that exhibits conchoidal fracture.

Concrete A thick liquid mixture of cement, aggregate, water, and other substances that sets into a hard, durable, solid material widely used for construction.

Conglomerate A clastic sedimentary rock composed of coarse, rounded fragments.

Contact Metamorphism Local metamorphism of preexisting rock caused by the intrusion of a nearby body of magma or hydrothermal fluids.

Continental Crust The part of the Earth's crust that forms the continents. It is largely composed of felsic rock types.

Continental Shelf An extension of a continental margin covered by relatively shallow salt water.

Convergent Plate Boundary A zone where two plates move toward each other.

Core The center zone of the Earth's interior, about 3,500 km (2,200 mi) in diameter, just below the mantle. It is composed of a liquid iron-nickel outer core and a solid iron-nickel inner core.

Corinthian Order The classical architectural order that features relatively slender columns with fluted shafts and ornate capitals decorated with acanthus leaves.

Course A horizontal row of bricks or dimension stone.

Craton The stable interior of a continent.

Crazing A network of fine cracks that develops on the surface of a glazed, burnt-clay product.

Cretaceous Period The span of geologic time from 145 to 66 Ma.

Crossbedding Patterns of curving planes or traces found within the strata of sandstone and other sedimentary rocks. Crossbedding indicates the general direction and force of the wind or water that originally laid down the sediments.

Crust The uppermost section of the solid Earth, ranging in thickness from about 6 to 50 km (4 to 30 mi). It lies directly above the mantle and directly below the atmosphere or oceans.

Crystalline Rock A general term for any igneous rock or metamorphic rock of igneous parentage.

Cyanobacterium (Plural Cyanobacteria) A relatively large and advanced photosynthetic bacterium. A member of the kingdom Eubacteria, it is sometimes inaccurately called a "blue-green alga."

Damp Course A base course made of nonabsorbent stone or another material that resists dampness and the uptake of ground moisture.

Dentil A somewhat tooth-shaped block or unit used for building ornament.

Devonian Period The span of geologic time from 419 to 359 Ma.

Diabase A phaneritic, intrusive, and mafic igneous rock. Diabase usually forms at shallow depths, in dikes and sills; its crystals, while still visible to the naked eye, are smaller than those of gabbro.

Dike A narrow, often vertical and wall-like mass of intrusive rock that forms when upward-moving magma cuts across preexisting rock bodies.

Dimension Stone Any quarried rock product that is cut to a specific size or shape for architectural or construction uses.

Distributary One of the branching channels of a stream flowing across a delta.

Divergent Plate Boundary A zone, characterized by rifts and volcanic activity, where the older portions of two plates are moving away from one another.

Dolomite The carbonate mineral CaMg(CO$_3$)$_2$. Also, a term used to denote the rock type chiefly made up of this mineral.

Dolostone The less ambiguous term for the rock type chiefly composed of the mineral dolomite.

Doric Order The classical architectural order that features columns with fluted shafts and simple, round capitals.

Dressed Referring to stone that has been prepared for architectural use. This preparation includes creating an essentially plane exposed face and also often applying an ornamental or protective finish to the exposed face. The finish can be applied with mechanical tools, abrasives, or chemical treatments.

Drift A collective term for material deposited directly or indirectly by a glacier.

Dunite A type of peridotite with a composition that is at least 90 percent olivine. It forms in the upper mantle beneath oceanic crust.

Efflorescence A white <u>weathering</u> crust produced when salt solutions migrate through porous <u>stone</u> and <u>brick</u> and precipitate on the surface.

End Moraine A ridge of <u>till</u> deposited at the leading margin of a <u>glacier</u> when the rate of ice melting at the margin is matched by new ice moving up from the rear.

English Basement A basement floor that is partly below <u>grade</u> and partly above.

Eoarchean Era The span of geologic time from 4.0 to 3.6 Ga.

Eocene Epoch The span of geologic time from 56 to 34 Ma.

Eon The largest subdivision of the <u>geologic time scale</u>.

Epicontinental Referring to a sea that covers a portion of a continent.

Epidote A green <u>silicate mineral</u> associated with <u>low-grade metamorphic rocks</u>.

Epoch On the <u>geologic time scale</u>, the largest subdivision of a <u>period</u>.

Era On the <u>geologic time scale</u>, the largest subdivision of an <u>eon</u>.

Erosion The process by which <u>sediments</u> and other Earth materials are removed by wind or running water.

Erratic A detached <u>rock</u> that had been transported by a <u>glacier</u> and subsequently deposited on the ground when the <u>glacier</u> melts.

Exedra A partially enclosed platform or room that features benches or other seating.

Exfoliation *See* Scaling.

Extrusive Referring to <u>igneous rock</u> formed by the cooling of <u>lava</u> or <u>tephra</u> on the Earth's surface (either on dry land or on the floor of a body of water).

Façade The front-facing portion of a building's exterior.

Face-Bedded Referring to <u>sedimentary rock</u> in a <u>quarry</u> that is cut on the same plane (parallel to) its <u>bedding</u>. If it is then used for <u>cladding</u> on a building and mounted perpendicular to the ground, it can prove more liable to <u>weathering</u> and <u>scaling</u> than <u>naturally bedded stone</u>.

Face Brick <u>Brick</u> of higher quality, hardness, durability, as well as greater ornamental appeal, that is used for building façades and interior walls intended for display.

Faience A term for glazed ceramic tile used for architectural ornament.

Fault A fracture in the <u>crust</u> where there has been significant displacement between the two sides.

Feldspar A term for any member of the very prevalent complex of <u>silicate minerals</u> that contain, among various other elements, aluminum, silicon, and oxygen.

Felsic Referring to <u>rocks</u> rich in silicon and aluminum. Such rocks include <u>granite, syenite, tonalite, quartz monzonite, granodiorite</u>, and <u>rhyolite</u>.

Ferromagnesian Mineral A <u>silicate mineral</u>, usually dark-colored, that contains a significant amount of iron and magnesium.

Ferruginous Containing iron oxide, or rust-colored.

Fieldstone (Architecture) Referring to an architectural style that features <u>fieldstones</u> set into the <u>façade</u> of a building.

Fieldstone (Geology) A detached <u>cobble</u> or <u>boulder</u> (in our area, usually an <u>erratic</u>) that is collected from farm fields and other open ground for architectural and construction purposes.

Fine-Grained Referring to <u>igneous</u> or <u>metamorphic stone</u> types whose <u>feldspar</u> crystals are less than 0.5 centimeter in length.

Finish The texture and surface appearance of a <u>stone</u> after its treatment by mechanical or chemical means. a finish is applied to either enhance a <u>stone's</u> aesthetic appeal or to better preserve it from weathering and wear.

Fissile Referring to a <u>rock</u> than can be readily split into sheetlike or slablike sections.

Flagging The setting of <u>flagstones</u> to create a durable, flat surface.

Flagstone A flat, thin-bedded <u>stone</u> of sufficient hardness and durability to be used as a path, sidewalk, patio, or roadway surface.

Flamed Finish Synonymous with Thermal Finish.

Flemish Bond A type of brickwork which consists of <u>courses</u> of alternating <u>headers</u> and <u>stretchers</u>.

Floating Foundation A raftlike structure, set in unconsolidated <u>sediments</u>, on which a building's load-bearing <u>piers</u> or walls are constructed. In Chicago, this construction technique generally predates the time when larger buildings were more effectively anchored with <u>piles</u> or <u>caissons</u>.

Flow An outpouring of <u>lava</u> on the Earth's surface.

Flow Banding A banded pattern in <u>igneous rock</u> usually caused by the shearing of viscous <u>magma</u> or <u>lava</u> when it comes into contact with a solid surface.

Fluting Decorative vertical grooves encircling the shaft of a <u>column</u>.

Foliated A term applied to <u>metamorphic rocks</u> that have <u>minerals</u> in a parallel alignment that often gives them a banded, wavy, or thinly layered appearance.

Footing A type of shallow building foundation. In modern times usually made of <u>concrete</u> containing steel reinforcing bars (rebars), it can be a widened strip set under an external wall, an individual pad, or entire raft set under the structure.

Foraminifer or Foram A type of planktonic marine organism, usually <u>microscopic</u> or almost so, that grows a <u>calcareous</u> test ("shell").

Fossil A preserved part or indication of an ancient organism's form or behavior.

Fossiliferous Containing <u>fossils</u>.

Freestone A quarryman's term for a <u>sedimentary rock</u> that can be easily sawn or cut in any direction without unwanted splitting, and thus is easily worked.

Fret An architectural design unit composed of a meander or another linear pattern. When set together, frets form one kind of classically inspired running ornament.

Frost Wedging A form of <u>weathering</u> in which liquid water fills cracks or pits in <u>stone</u> or pavement and then expands upon freezing, causing cracking or <u>spalling</u>.

Gabbro A <u>phaneritic</u>, dark-colored, and quartz-poor <u>mafic</u> <u>igneous rock</u>. It is the <u>intrusive</u> equivalent of <u>basalt</u>.

Gable The triangular portion of a building between two slopes of the roof that meet at their top.

Ganderia A <u>microcontinent</u> that collided with <u>Laurentia</u> in the interval between the <u>Taconic Orogeny</u> and the <u>Acadian Orogeny</u>. It has been hypothesized that Ganderia originally broke off the margin of <u>Gondwana</u>.

Geologic Time Scale The visual representation, often drawn to scale and displayed as a horizontal or vertical bar, of the Earth's history, from its origin approximately 4.6 Ga to the present. It is divided into these subdivisions of decreasing order and size: <u>eons</u>, <u>eras</u>, <u>periods</u>, and <u>epochs</u>.

Glaciation An episode (often lasting about 100 ka) consisting of the formation, advance, and retreat of one continental ice sheet. Not synonymous with ice age (which see).

Glacier A large, persistent mass of ice that moves outward under the weight of its own thick center on low slopes, or downward on steep valley slopes. A glacier forms when there is, for an extended period, more net annual snowfall than net annual melting.

Gneiss A high-grade <u>metamorphic rock</u> characterized by linear bands of dark and light <u>minerals</u>.

Gneissic A term applied to <u>granites</u> and other similar <u>igneous rocks</u> that show some evidence of banding or <u>foliation</u> suggestive of some degree of metamorphism.

Gondwana or Gondwanaland The <u>Paleozoic-era supercontinent</u> that included Africa, India, South America, Australia, and Antarctica.

Grade The current surface level; ground level.

Grade Course A <u>course</u> of brick or <u>stone</u>, often the bottommost on a building exterior, at <u>grade</u>.

Graded Bed A bed of sediments that changes, from bottom to top, from coarser particles to finer particles.

Grainstone A type of carbonate rock in the Dunham classification system that is composed of grains that are not surrounded by a mud matrix.

Grand Art Deco Formula A term coined by the author to describe a common design pattern found in skyscrapers and other buildings embodying the Art Deco style. It consists of a massive, uniform Salem Limestone exterior contrasted at its base by at least somewhat darker, crystalline rock.

Granite (Architecture) An imprecise and variably defined term that generally refers to any hard, silicate rock type (be it igneous, sedimentary, or metamorphic) that can take a high polish.

Granite (Geology) A phaneritic, usually light-colored, felsic, igneous rock. The intrusive equivalent of rhyolite, it contains at least 10 percent quartz.

Granitoid Referring to the group of igneous intrusive rocks that includes granite and such closely related types as alkali-feldspar granite, granodiorite, syenite, quartz monzonite, aplite, tonalite, and trondhjemite.

Granodiorite A granitoid rock that has 65 to 95 percent of its feldspar content in the form of plagioclase, and is 20 to 60 percent quartz.

Gravel A general term for rock particles of pebble size.

Grenville Front The tectonic zone in eastern Canada that is the boundary between Laurentia's older, inner portion and the outer section added during the Grenville Orogeny.

Grenville Orogeny The widespread and multiphased orogeny that occurred ca. 1.3 Ga to 980 Ma.

Grillage A type of shallow building foundation made of a network of crisscrossed timber beams or iron or steel rails.

Groundmass The matrix of small crystals in an igneous rock in which phenocrysts are set.

Groundwater Water held beneath the Earth's surface in rock, sediments, or the soil.

Gypsum The sulfate mineral $CaSO_4 \cdot 2H_2O$.

Hadean Eon The span of geologic time from 4.6 to 4.0 Ga.

Halide Mineral A member of a chemical group of minerals that contain, among other substances, a halide element (fluorine, chlorine, bromine, etc.).

Halite The halide mineral NaCl. In its rock form it is also known as "rock salt."

Hardpan A zone of soil or unconsolidated sediments that has been compacted into a stiffer, rocklike consistency. In Chicago, this term should for clarity's sake refer specifically to a zone of stiff, dewatered till approximately 50 to 85 feet below the modern surface. However, architects and historians arbitrarily use it for any harder layer they happen to be describing at the moment.

Hardpan Caisson. A deep caisson that extends downward to the hardpan layer above the bedrock.

Header A brick or block of dimension stone in ashlar masonry that is set with its short face exposed.

Headwall The vertical working face of a quarry, from which stone is extracted.

Hematite The oxide mineral Fe_2O_3, a principal iron ore.

High-Grade Metamorphic Rock A metamorphic rock that has been subjected to a relatively high increase in pressure and to temperatures above 320° Celsius.

Honed Finish A finish created by treating a stone surface with abrasives in a way that results in a nonreflective surface and somewhat duller colors.

Hornblende A common, black, rock-forming silicate mineral with the general formula of $(Ca,Na)_{2-3}(Mg,Fe,Al)_5(Si,Al)_8O_{22}(OH,F)_2$.

Hot Spot The surface expression of a mantle plume. Hots spots are characterized by dramatic volcanic activity and massive lava flows.

Hydrothermal Fluids Hot and chemically reactive fluids circulating underground.

Ice Age A major, extended event in Earth history (often lasting 10 Ma or more) that includes at least several glaciations and interglacials.

Igneous Referring to any rock formed by the cooling of magma, either underground or at the Earth's surface.

Ignimbrite A rock formed from the ash and other materials deposited by a pyroclastic flow.

Ilmenite An oxide mineral with the chemical formula $FeTiO_3$.

Interreef Referring to rock strata that form from sediments that are deposited between or at some distance from reefs.

Intrusive Referring to igneous rock formed from magma cooling beneath the Earth's surface.

Ionic Order The classical architectural order that features columns that are usually fluted and have capitals with volutes.

Island Arc A curving line of volcanic islands that forms over a subducting plate.

Isotope A form of an element that has atoms with a nonstandard number of neutrons.

Isotopic Age *See* Radiometric Age.

Joint (Architecture) The gap or space between dimension stones or bricks, or the edge of a dimension stone or brick.

Joint (Geology) A fracture in bedrock where there has been no significant displacement between the two sides.

Jurassic Period The span of geologic time from 201 to 145 Ma.

Kaskaskia Sequence The sequence dating from the early Devonian period to the early Pennsylvanian subperiod.

Komatiite A rare extrusive igneous rock that is the ultramafic equivalent of basalt.

Labradorescence A property of some minerals that have an internal structure composed of a plane of molecules that reflects light to produce iridescent colors. Also referred to as schiller.

Lagging As used in Chicago architecture and engineering, wood slats used to line, stabilize, and seal the sides of caissons before they are filled with concrete.

Landform Any distinct feature on a planet's surface produced by natural forces.

Lapilli Tephra particles that have a diameter of 2 to 64 mm (0.1 to 2.5 in). In other words, they're larger than ash particles.

Lapilli Tuff An igneous, extrusive rock composed of lapilli that are welded or cemented together.

Laterite A clayey material, often reddish or orange in color, that is derived from heavily weathered rock and is characteristic of wet, tropical climates. It contains high concentrations of aluminum and iron compounds.

Laurentia The ancestral version of North America as it existed in the Proterozoic eon and Paleozoic era.

Lava Molten rock at the Earth's surface (either on land or on the ocean bottom).

Leucogranite A granite that is light-colored due to an absence of substantial dark-hued mineral content. Leucogranites are thought to be the product of continental collision; their magmas are derived from metamorphosed sedimentary rocks found in the upper crust.

Liesegang Ring A linear pattern, often curving, wavy, or ringlike, produced by iron-oxide minerals precipitating out of groundwater in sandstone or other porous rock.

Lime A substance containing mainly calcium oxide (CaO) or calcium hydroxide ($Ca(OH)_2$).

Limestone A <u>sedimentary rock</u> chiefly composed of the <u>mineral calcite</u>. It is often <u>chemically precipitated</u>, but it may <u>clastic</u> instead.

Lithified Referring to <u>sediments</u> that have been turned into solid <u>rock</u> through compaction or natural cementation.

Lithology The description of a <u>rock's</u> or rock unit's characteristics and identification traits.

Lithosphere The brittle, solid uppermost section of the solid Earth, comprising the <u>crust</u> and uppermost <u>mantle</u>.

Longshore Current The nearshore movement of water and water-borne <u>sediments</u> parallel to the coastline.

Longshore Drift The in-and-out motion of the surf and surf-borne <u>sediments</u> that produces a zigzag pattern of movement down the coastline.

Lower When applied to geologic time units, it denotes the oldest or earliest portion of that unit. For example, Lower <u>Jurassic</u> refers to the earliest portion of the <u>Jurassic period</u>.

Low-Grade Metamorphic Rock A <u>metamorphic rock</u> that has been subjected to a relatively gentle increase in pressure and to temperatures between 200 and 320 degrees Celsius.

Macroscopic Visible to the naked eye; not requiring magnification.

Mafic Referring to dark-colored <u>igneous rocks</u> that are rich in iron and magnesium. Such rocks include <u>basalt</u> and <u>gabbro</u>.

Magma Molten rock below the Earth's surface.

Magnesite A <u>carbonate mineral</u> with the chemical formula $MgCO_3$.

Magnetite An <u>oxide mineral</u> with the chemical formula Fe_3O_4. It is one of the main <u>ores</u> of iron and is well known for its magnetic properties.

Mansard Roof A roof that features two different slopes, with the lower being steeper than the upper.

Mantle The zone of the Earth's interior, about 1800 mi deep, that lies directly under the <u>crust</u>.

Mantle Plume A large column of superheated <u>magma</u> that rises through the <u>mantle</u> and, if it reaches the surface, creates a <u>hot spot</u>.

Marble (Architecture) An imprecise and variably defined term that generally refers to any <u>carbonate</u> and some noncarbonate <u>rock</u> types that are relatively soft and can take a high polish. Examples include true <u>marble</u>, <u>limestone</u>, <u>serpentinite</u>, <u>breccia</u>, and <u>ophicalcite</u>.

Marble (Geology) The <u>metamorphic</u> equivalent of such <u>carbonate rocks</u> as <u>limestone</u> and <u>dolostone</u>. It has been subjected to high temperatures, and is widely used in sculpture and <u>cladding</u> because it takes a high polish.

Marshfield Terrane The <u>microcontinent</u> that collided with the <u>Superior Craton</u> during the final phase of the <u>Penokean Orogeny</u>.

Masonry The use of <u>dimension stone</u>, <u>brick</u>, <u>concrete</u> blocks, or other hard materials for wall and building construction.

Medium-Grained Referring to <u>igneous</u> or <u>metamorphic stone</u> types whose <u>feldspar</u> crystals are between 0.5 and 1.0 centimeter in length.

Mesoarchean Era The span of geologic time from 3.2 to 2.8 Ga.

Mesoproterozoic Era The span of geologic time from 1.6 to 1.0 Ga.

Mesozoic Era The span of geologic time from 252 to 66 Ma.

Metamorphic Referring to any <u>rock</u> that has been subjected to increased temperature, increased pressure, or both, and therefore has been transformed into a type different than its original form.

Metanorthosite An anorthosite that has undergone metamorphism.

Metaquartzite A quartzite of metamorphic origin.

Meteorite An object of interplanetary origin that reaches the surface of the Earth without completely burning up in the atmosphere.

Mica A term for the family of rock-forming silicate minerals that exhibit flat, sheetlike crystalline structure.

Microcline An alkali feldspar mineral of the composition $KAlSi_3O_8$.

Microcontinent A detached section of continental crust that is smaller than a full continent.

Microcrystalline Referring to crystals that are too small to be seen with the naked eye.

Microscopic Not visible to the naked eye; requiring magnification.

Midcontinent Rift The large, arc-shaped rift situated from Oklahoma through the Lake Superior region to Alabama that formed in the Mesoproterozoic era, from approximately 1.1 to 1.0 Ga. This rift might have resulted in the breakup of Laurentia had it not been for the crustal compression generated by the Grenville Orogeny, though new evidence makes this uncertain.

Migmatite A rock that is considered a hybrid type because it is composed of both igneous and metamorphic components. One example would be a gneiss that contains veins or other inclusions of granite.

Mineral A naturally occurring, inorganic substance that has a definite chemical composition. Many minerals form crystals.

Mineralogy The study of rock-forming minerals.

Miocene Epoch The span of geologic time from 23 to 5 Ma.

Mississippian Subperiod The subdivision of the Carboniferous period from 359 to 323 Ma.

Mollusc An invertebrate animal of the phylum Mollusca. Molluscs include snails and cephalopods.

Monzogranite A granite that has between 35 and 65 percent of its total feldspar content in the form of plagioclase.

Monzonite A granitoid rock containing approximately equal amounts of plagioclase and alkali feldspars, and 5 percent or less of quartz.

Moraine See End Moraine.

Mortar A substance used by masons to bind dimension stones or bricks together at their joints. While the composition of mortars has varied over the centuries, modern mortar characteristically contains Portland Cement, sand, extra lime, and water.

Muscovite A light-colored, sheet-forming silicate mineral of the mica family. One of its listed chemical formulas is $KAl_2(Si_3Al)O_{10}(OH,F)_2$; other variations are cited.

Narthex That portion of a church interior just inside the entranceway or vestibule.

Naturally Bedded Referring to sedimentary rock in a quarry that is cut perpendicular to its bedding.

Nautiloid A member of an ancient line of cephalopods that was most prevalent in the Paleozoic era. One surviving species of this group is the chambered nautilus.

Nave The main interior section of a church between the apse and the narthex.

Negative Feedback Loop A process in which an initial condition or effect triggers its own mitigation or diminishment.

Neoarchean Era The span of geologic time from 2.8 to 2.5 Ga.

Neoclassical Referring to the architectural style employed from the Renaissance to modern times that is based on the architectural orders and elements of ancient Greece and Rome.

Neoproterozoic Era The span of geologic time from 1.0 Ga to 541 Ma.

Nodular Containing nodules or having a lumpy or blotchy appearance.

Nonfoliated A term applied to metamorphic rocks that do not have minerals in a parallel alignment. As a result, they do not have a banded, wavy, or thinly layered appearance.

Nummulites A genus of giant foraminifers.

Oceanic Crust The part of the Earth's crust that floors ocean basins. This crust is largely composed of mafic rocks.

Oligocene Epoch The span of geologic time from 34 to 23 Ma.

Olivine A silicate mineral notable for its green color. Its chemical formula is (Mg^{2+}, Fe^{2+})$_2$SiO.

Onyx A form of the mineral chalcedony characterized by light and dark bands.

Oolitic A term referring to a limestone composed of tiny, rounded grains of calcite.

Ophicalcite A recrystallized metamorphic rock containing carbonate minerals and brecciated serpentinite.

Ordovician Period The span of geologic time from 485 to 444 Ma.

Ore A mineral or rock type, usually metallic, that can be mined at a profit.

Orogenesis The process of mountain building.

Orogeny A mountain-building event.

Orthoclase A common alkali-feldspar mineral: K($AlSi_3O_8$).

Orthoquartzite A quartzite of sedimentary origin that has not undergone metamorphism.

Outcrop An exposure of bedrock.

Outwash A deposit of sorted sand, pebbles, and cobbles carried and then deposited by glacial meltwater streams.

Overburden Any loose material that sits atop the bedrock: soil, unconsolidated sediments, etc. Also, any bedrock that lies over a desired vein of coal or ore.

Oxide Mineral A member of a chemical group of minerals that contain, among other substances, O_2.

Paleoarchean Era The span of geologic time from 3.6 to 3.2 Ga.

Paleocene Epoch The span of geologic time from 66 to 56 Ma.

Paleoproterozoic Era The span of geologic time from 2.5 to 1.6 Ga.

Paleozoic Era The span of geologic time from 541 to 252 Ma.

Palmette A type of classical ornament that resembles a stylized palm frond.

Pangaea (sometimes rendered Pangea) The supercontinent that existed from approximately 300 to 200 Ma.

Parent Rock The original rock type from which a metamorphic rock forms.

Pebble A rock particle 2 to 64 mm (.08 to 2.5 in) in diameter.

Pediment The usually broad gable or squat triangular element, very common in classical architecture, set above the entablature.

Pegmatite A very coarse-grained granite.

Pembine-Wausau Terrane The island arc that collided with the Superior Craton during the initial phase of the Penokean Orogeny.

Pennsylvanian Subperiod The subdivision of the Carboniferous period from 323 to 299 Ma.

Penokean Orogeny A Paleoproterozoic mountain-building event affecting what is now Wisconsin. It was triggered by collisions of the Superior Craton with the Pembine-Wausau Terrane and then Marshfield Terrane.

Peridotite A dense, dark-toned, ultramafic rock thought to originate in the Earth's mantle.

Period On the geologic time scale, the largest subdivision of an era.

Permian Period The span of geologic time from 299 to 252 Ma.

Perthite A silicate mineral composed of the intergrowth of two different types of feldspar.

Petrology The study of a <u>rock's</u> chemical properties, mineralogical content, and environment of formation.

Phaneritic Composed a <u>macroscopic</u> crystals.

Phenocryst A large crystal in an <u>igneous rock</u> that is set in a <u>groundmass</u>.

Phosphate Mineral A member of a chemical group of <u>minerals</u> that contain, among other substances, PO_4.

Pier A <u>column</u> or other vertical element that bears a structural load.

Pilaster A flat-faced <u>column</u> attached to the wall behind it. It is an ornamental rather than a load-bearing element.

Pile A pole or <u>column</u> driven into the <u>substrate</u> to provide support and stability for a structure above it.

Pit An open-air facility where unlithified <u>sediments</u> (sand, gravel, etc.) are mined.

Pitch The sloped surface of a roof.

Plagioclase Feldspars One of the two main groups of <u>feldspar minerals</u>. Its members, which form a continuous series, are albite, oligoclase, andesine, labradorite, bytownite, and anorthite. Plagioclase feldspars are sometimes called "soda-lime feldspars" instead.

Plaster A substance composed of <u>lime</u>, <u>gypsum</u>, or <u>cement</u> mixed with water that is applied to walls and other surfaces. Once dried, it forms a decorative or protective coating.

Plate One of about fifteen rigid sections that make up the Earth's <u>lithosphere</u>. A plate can contain <u>oceanic crust</u> only, or both oceanic and <u>continental crust</u>.

Plate Tectonics The theory first developed in the middle of the twentieth century that explains various geological phenomena in the context of moving <u>plates</u>.

Pleistocene Epoch The span of geologic time from 2.6 Ma to 12 ka.

Plinth The base of a pedestal, column, or statue. Also, the basal part of a building's exterior.

Plinth Courses The lowermost courses of stone on a building's exterior, especially if they have the appearance of being a distinct base for the building.

Pliocene Epoch The span of geologic time from 5 to 2.6 Ma.

Pluton A mass of <u>intrusive igneous rock</u> that was originally emplaced underground.

Polished Finish A <u>finish</u> created by treating a <u>stone</u> surface with abrasives in a way that produces a highly reflective surface that emphasizes the <u>stone's</u> natural colors.

Porphyritic Referring to an <u>igneous rock</u> that has <u>phenocrysts</u> set in a <u>groundmass</u>.

Porphyry An <u>igneous rock</u> containing large <u>phenocrysts</u> embedded in a <u>groundmass</u>.

Porte Cochère Literally, a "carriage entrance"; a covered entryway.

Portland Cement The most frequently used modern <u>cement</u> used for construction practices. It is produced by firing <u>limestone</u> and <u>clay</u>, grinding the product, called clinker, into a powder, and adding to it a small amount of <u>gypsum</u>.

Positive Feedback Loop A process in which an initial condition or effect triggers its own further increase or intensification.

Potassium Feldspars The subgroup of the <u>alkali feldspars</u> that comprises <u>orthoclase</u>, <u>microcline</u>, and sanidine.

Pozzolana Volcanic <u>ash</u>, originally mined by the ancient Romans in the Pozzuoli, Italy area, that is used in the formulation of some types of <u>cement</u>.

Proterozoic Eon The span of geologic time from 2.5 Ga to 541 Ma.

Ptygmatic Folds Tight and often asymmetrical folds, frequently of varying sizes, in <u>rock</u>. They have a much greater amplitude (height of fold from trough to crest) than wavelength (width of fold from trough to trough). Ptygmatic folds usually form when the material of the folded layer is much stiffer or more viscous than the surrounding material.

Pyroclastic Flow In a general sense, an extremely hot cloud of particles and gas that move down a volcano's slope at up to 600 mph. In a more restricted sense, it is a hot cloud of particles and gas that has laminar flow, i.e., it moves downhill as a ground-hugging sheet.

Pyroxene A term referring to a group of chemically related and often dark-colored silicate minerals.

Quarry An open-air facility or enclosed-shaft mine where the local bedrock is extracted.

Quarry-Faced Stone Essentially synonymous with rock-faced stone.

Quartz An extremely prevalent mineral composed of silicon and oxygen. Classified as either a silicate or oxide mineral, its chemical formula is SiO_2.

Quartzite Either a metamorphic equivalent of sandstone (metaquartzite), or an unmetamorphosed sandstone with grains cemented with silica (orthoquartzite).

Quartz Monzonite A granitoid rock containing approximately equal amounts of plagioclase and alkali feldspars, and between 5 and 20 percent quartz.

Quaternary Period The span of geologic time from 2.6 Ma to the present.

Queen Anne Style A style of largely domestic architecture noted for its eclectic blend of other earlier styles and its wide variety of building materials.

Quoin The masonry that forms the corner of a building.

Radiometric Age, Radiometric Date The age of a rock or other geologic specimen determined by measuring the amount of decay of an unstable isotope it contains.

Rapakivi A term from the Finnish language for igneous rock containing large and usually pink potassium-feldspar crystals that are bordered by white, green, or black plagioclase-feldspar or alkali-feldspar zones.

Rebar A contraction of "reinforcing bar," as in the steel bars set in concrete to strengthen it.

Reduction Zone A pale-colored, spherical or oblong zone in sandstone where the oxidation of hematite has been inhibited by an enclosed pebble or organic matter.

Reef A term with various definitions, but in paleontology and geology it is defined as a marine structure largely made of lime-secreting organisms.

Reefal Referring to rock formed in a reef environment.

Regional Metamorphism The metamorphism of preexisting rock caused by tectonic activity acting over a large area.

Rhyolite A quartz-rich, felsic, aphanitic, and usually light-colored igneous rock. The extrusive equivalent of granite, it is at least 10 percent quartz.

Richardsonian Romanesque An architectural style in favor in the late nineteenth century that features such medieval Romanesque elements as semicircular arches, rock-faced stone surfaces, conical towers, and a massive, fortresslike appearance. Named for its formulator, the American architect Henry Hobson Richardson (1838–1886), it also often employs somber, dark-toned brownstone exteriors.

Ripple Marks Rippled patterns, originally made in sediments by flowing water, surf action, or the wind, that have become lithified and preserved in sandstone and other sedimentary rocks.

Riprap Large rock fragments used to stabilize a shoreline or to help stabilize a slope or hillside.

Rock A consolidated assemblage of one or more types of minerals.

Rock Caisson. A deep caisson that extends downward all the way to the bedrock.

Rock-Faced Stone A form of dimension stone in which the joints of the stone are chiseled away, but the central portion is left with a jagged, rough, projecting surface.

Rockhound A superior sort of person; one who recognizes that the collecting, scientific study, and aesthetic appreciation of rock types is a primal, necessary, and transcendent human activity. Not to be confused with those credulous souls who keep "pet rocks" in little cardboard houses or stake their wellbeing on "mystical healing crystals."

Rodinia The supercontinent that included most modern continents and that existed from approximately 1.1 Ga to 750 Ma.

Roman Brick Brick that resembles the long, shallow-flan brick type used by the ancient Romans. Its dimensions are often approximately 2 x 4 x 12 inches (5 x 10 x 30 centimeters). Especially favored by such Prairie Style architects as Sullivan, Wright, and Elmslie, it tends to emphasize a building's horizontal lines.

Romanesque A medieval architectural style that characteristically features such elements as semicircular arches, flamboyant columns, massive towers, and barrel vaults.

Rubbed Finish A finish that is produced by rubbing the stone surface with sand or some other hard substance.

Rubble Architectural stone that has not been "squared off" (cut with sides meeting at right angles). Rubble walls are usually uncoursed.

Rusticated Stone Stonework in which the joints are recessed or chamfered, so that the rest of the stone surface projects outward.

Saline Containing salt—often specifically the mineral halite.

Sand A rock or mineral particle .06 to 2 mm (.0024 to .08 in) in diameter.

Sandstone A clastic rock composed of sand grains that have been cemented together with silica, calcite, or another mineral.

Scagliola An ornamental, polished plaster surface composed of gypsum, isinglass, alum, and coloring agents. It is used by architects as a relatively inexpensive stand-in for marble.

Scaling Weathered pieces that flake or peel off a stone surface.

Seam-Faced Stone A form of dimension stone in which the side of the stone that was already naturally weathered before it was quarried is used as the external face on a building. This gives the wall or façade a nicely aged and venerable appearance.

Seasoning A quarryman's term for the process of drying out porous rock such as sandstone, especially if it is extracted from below the water table. This is usually accomplished by just letting the stone stand for a prolonged period in a dry place.

Second Empire Style An architectural style current in the nineteenth century that features among other characteristic elements mansard roofs.

Sediment Unconsolidated material made of rock or mineral fragments.

Sedimentary Referring to any layered rock type formed either by clastic particles or by chemical precipitation. Examples include breccia, chert, conglomerate, dolostone, limestone, sandstone, and shale.

Sequence A stratigraphic term referring to a large assemblage of marine sedimentary rocks bounded above and below by unconformities. Such a deposit represents a span of geologic time when sea level rose and saltwater seas covered a large portion of continental interiors.

Serpentine A term for a group of silicate minerals that are often pale to dark green. They are usually produced in metamorphism.

Serpentinite An exotic, serpentine-containing metamorphic rock whose parent rock is dunite. Usually deep-green with snaking white veins, this rock is thought to originate near the Earth's crust-mantle boundary.

Shale A clastic sedimentary rock composed of clay-sized particles.

Silica Synonymous with Quartz.

Silicate Mineral A member of a chemical group of minerals that contains, among other substances, SiO_4. The one exception, quartz, is SiO_2.

Siliceous Containing silica.

Sill (Architecture) The shelflike, horizontal unit at the base of a window or door.

Sill (Geology) A horizontal or slanting body of intrusive rock that forms when upward-moving magma pushes its way between two strata.

Silt A <u>mineral</u> particle 0.003 to 0.06 mm (0.0001 to 0.003 in) in diameter.

Siltstone A <u>clastic</u>, <u>sedimentary rock</u> composed of <u>silt</u> particles.

Silurian Period The span of geologic time from 444 to 419 Ma.

Slate The <u>foliated</u>, <u>low-grade metamorphic</u> equivalent of <u>shale</u>.

Spall To flake off, fall off, or splinter from a mass of <u>rock</u>.

Spandrel A horizontal panel that separates the windows of one story from those of the story above or below it.

Spread Foundations A network of separate, detached mats or flat bases set in the shallow <u>substrate</u> to which a building's vertical supports are attached.

Staff A building material, usually used for temporary structures, composed of plaster of Paris, hemp or other plant fiber, and <u>cement</u>.

Statuario A term for the highly prized variety of <u>marble</u> that has little or no veining and is pure white, or almost so. It is also more translucent than other varieties and has a texture that some sources describe as "waxy."

Stone (Architecture) Any <u>rock</u> type used for decorative purposes that cannot take a high polish.

Stone (Geology) Essentially synonymous with <u>rock</u>; it can refer to any <u>igneous</u>, <u>sedimentary</u>, or <u>metamorphic rock</u> type regardless of its properties, origin, or architectural use.

Stratigraphy The subdiscipline of geology that is the study, classification, and dating of <u>rock strata</u>.

Stratum (Plural = Strata) A layer or bed of <u>sedimentary rock</u>.

Stretcher A block of <u>brick</u> or <u>dimension stone</u> in <u>ashlar</u> masonry that is set with its long face exposed.

Stringcourse A horizontal <u>course</u> of <u>masonry</u> on a wall that is usually quite distinct from what lies below and above it.

Stylolite A wavy or jagged linear pattern found in <u>limestone</u> and other <u>rock</u> types. It contrasts in color with the surrounding <u>rock</u> and is composed of <u>mineral</u> or organic matter. Stylolites are often considered an ornamental asset in decorative <u>stone</u>.

Subduction The process in which one <u>plate</u> sinks beneath another.

Substrate The soil, <u>sediment</u>, or <u>rock</u> that lies directly under the Earth's surface.

Sugared Denoting the rough, granular, uneven or pitted surface of badly <u>weathered</u> <u>marble</u>.

Sulfide Mineral A member of a chemical group of <u>minerals</u> that contain, among other substances, -S (sulfur).

Supercontinent A giant landmass formed by the collision of two or more continents.

Superior Craton One of the <u>Archean</u> crustal sections that formed the ancient core of <u>Laurentia</u>. It is located in central Canada and in the uppermost part of North-Central US.

Syenite A <u>felsic</u> and <u>intrusive igneous rock</u> similar in appearance to <u>granite</u> but with significantly less <u>quartz</u> content.

Taconic (or Taconian) Orogeny The <u>orogeny</u> that occurred in the <u>Ordovician</u> and early <u>Silurian periods</u> when a volcanic-island arc collided with <u>Laurentia</u>.

Tectonic Referring to the movement, deformation, or structural changes of the Earth's <u>crust</u>.

Tephra A general term for any airborne particles or fragments ejected from a volcano, regardless of their size.

Terra-Cotta Molded and fired <u>clay</u> used as architectural ornament (in glazed form) and for roofing tiles (either unglazed or glazed).

Terrain The surface expression of a landscape; the "lie of the land."

Terrane A three-dimensional section of the Earth's crust that is bounded by faults on all sides and that is composed of rock units that share a common origin.

Tertiary Period The span of geologic time from 66 to 2.6 Ma.

Tethys Ocean (Tethys Sea, Tethys Seaway) A major body of water, floored by oceanic crust, that formed in the Mesozoic era. Its much smaller modern remnants are the Mediterranean, Black, Caspian, and Aral Seas.

Thermal Finish A finish created by treating a stone surface with a high-temperature flame from a blowpipe. This produces a rough, nonskid texture.

Till The form of drift that is deposited either under or directly in front of a glacier. It is composed of all sediment sizes, from clay and silt to gravel and boulders. For that reason it is considered unsorted drift.

Tonalite A felsic or intermediate, intrusive igneous rock that has its feldspar content mostly in the form of plagioclase feldspar, with 10 percent or less alkali feldspar; also, its quartz content is at least 20 percent. It is thought that tonalites form from the melting of mafic ocean crust when it is subducted at a convergent plate margin.

Tooled Finish A finish created by treating a stone surface with a large, single-pointed hammering tool. The result is similar to a bush-hammered finish, but it can be selectively applied to a small area.

Transept The section of a church that lies perpendicular to the main axis to form the two arms of the cross.

Travertine A limestone formed by the deposition of calcite or other carbonate minerals by geothermally heated spring water at normal temperature. Travertine often forms at hot springs.

Triassic Period The span of geologic time from 252 to 251 Ma.

Trim An architectural term for stone, brick, or terra-cotta used as a highlighting contrast to the main material of a building's exterior. Trim often takes the form of thin horizontal courses, window sills, or capstones on church buttresses.

Troctolite A phaneritic, mafic, igneous rock composed mainly of olivine and calcium-rich plagioclase feldspar. It resembles gabbro but lacks the latter's higher pyroxene content.

Trondhjemite A type of tonalite that specifically has oligoclase as its plagioclase feldspar component.

Tuff An igneous rock formed from volcanic ash.

Turbidite Deposit A deposit of sediments laid down underwater by a turbidity current.

Turbidity Current A fast-moving current of dense, sediment-laden water. It can be caused by an earthquake or underwater avalanche.

Tuscan Order The classical architectural order that is in effect a simplified Doric order. It features columns that are not fluted.

Unconformity A gap in the stratigraphic record caused by an episode of erosion or no new formation of rock.

Uncoursed A term that refers to masonry that is set in a random pattern, and not in courses.

Underclay The layer of clay found directly beneath a stratum of Pennsylvanian (Upper Carboniferous) coal. Underclays originally formed as the soils in which the coal-producing vegetation was rooted.

Ultramafic Referring to igneous rocks that are very rich in iron and magnesium. They are rarely found in the Earth's crust but thought to be the main constituent of much of the Earth's mantle. They include peridotite and its variant dunite.

Upper When applied to geologic time units, it denotes the youngest or latest portion of that unit.

Variscan Orogeny The European mountain-building event that occurred in the late Carboniferous and Permian as part of the assembly of Pangaea.

Volute A decorative element of capitals of the Ionic order and the Composite order. It resembles the rolled-up end of a scroll.

Vug A hole or cavity in a rock.

Water Table (Architecture) A course of masonry, either at the base of a building's exterior or above it, that projects outward somewhat from the rest of the wall. A water table helps direct rain water away from the building's foundation, but is sometimes used by masons and architects as a purely decorative feature.

Water Table (Geology) The plane or surface that is the boundary between the zone of aeration and the zone of saturation.

Weathering The process by which rock, soil, and other materials exposed to the elements undergo chemical or physical change due to such factors as the weather, pollution, salt compounds, or the action of organisms. While weathering sometimes produces attractive and desirable coloration in building stone, it can in other situations be extremely destructive.

Well-Sorted Referring to a sediment in which the particles are of one uniform size. In engineering parlance, this term can mean the opposite (a wide range of particle sizes is present). But in this book the primary, geological definition is used.

Wisconsin Glaciation The final ice-sheet advance of the Pleistocene epoch. It lasted from 75 ka to 12 ka.

Yavapai Orogeny An orogeny that occurred at about 1.7 Ga when an island arc collided with the underside of Laurentia.

Selected Bibliography

GENERAL GEOLOGY

Chrzastowski, Michael J. *Chicagoland: Geology and the Making of a Metropolis; Field Excursion for the 2005 Annual Meeting Association of American State Geologists*. Champaign, IL: Illinois State Geological Survey, 2005.

Chrzastowski, Michael J. *The Chicago River—a Legacy of Glacial and Coastal Processes: Guidebook for the 2009 Meeting of the North-Central Section of the Geological Society of America*. Champaign, IL: Illinois State Geological Survey, 2009.

Dorr, John A., Jr., and Donald F. Eschman. *Geology of Michigan*. Ann Arbor, MI: University of Michigan Press, 1970.

Heiken, Grant, Renato Funiciello, and Donatella De Rita. *The Seven Hills of Rome: A Geological Tour of the Eternal City*. Princeton, NJ: Princeton University Press, 2005.

Higgins, Michael Denis, and Reynold Higgins. *A Geological Companion to Greece and the Aegean*. Ithaca, NY: Cornell University Press, 1996.

Illinois State Geological Survey. "ILSTRAT: The Online Handbook of Illinois Stratigraphy." Accessed September 3, 2019. http://isgs.illinois.edu/ilstrat/index.php/Main_Page.

LaBerge, Gene L. *Geology of the Lake Superior Region*. Phoenix, AZ: Geoscience Press, 1994.

Mikulic, Donald G., and Joanne Kluessendorf. *The Classic Silurian Reefs of the Chicago Area*. ISGS Guidebook 29. Champaign, IL: Illinois State Geological Survey, 1999.

Sampsell, Bonnie M. *The Geology of Egypt: A Traveler's Handbook*. Cairo: American University in Cairo, 2014.

United States Geological Survey. Mineral Resources Online Spatial Data (website). https://mrdata.usgs.gov.

Wiggers, Raymond. *Geology Underfoot in Illinois*. Missoula, MT: Mountain Press, 1997.

Willman, H. B., Elwood Atherton, T. C. Buschbach, Charles Collinson, John C. Frye, M. E. Hopkins, Jerry A. Lineback, and Jack A. Simon. *Handbook of Illinois Stratigraphy*. Urbana, IL: Illinois State Geological Survey, 1975.

Worthen, Amos Henry. *Economical Geology of Illinois*. Vols. 1–3. Springfield, IL: H. W. Rokker, 1882.

ARCHITECTURAL-STONE TYPES AND QUARRYING

Austin, Muriel B., Arthur M. Hussey II, and John R. Rand. *Maine Granite Quarries and Prospects*. Maine Geological Survey Mineral Resources Index, no. 2. Augusta, ME: Maine Department of Economic Development, 1958.

Bowles, Oliver. *The Stone Industries*. 2nd ed. New York: McGraw-Hill, 1939.

Bowles, Oliver. *The Structural and Ornamental Stones of Minnesota*. US Geological Survey Bulletin 663. Washington, DC: US Government Printing Office, 1918.

Bownocker, J. A. *Building Stones of Ohio*. Bulletin (Geological Survey of Ohio), 4th ser., no. 18. Columbus, OH: Geological Survey of Ohio, 1915.

Buckley, Ernest Robertson. *On the Building and Ornamental Stones of Wisconsin.* Madison, WI: Wisconsin Geological and Natural History Survey, 1898.

Buckley, Ernest Robertson, and Henry Andre Buehler. *The Quarrying Industry of Missouri.* Jefferson City, MO: Missouri Bureau of Geology and Mines, 1904.

Coldspring Corporation. "Granite in Chicago Buildings" (typewritten 1962 manuscript with later handwritten additions). In the archives of the Chicago History Museum, Chicago, IL.

Dale, T. Nelson. *The Commercial Granites of New England.* United States Geological Survey Bulletin 723. Washington, DC: US Government Printing Office, 1923.

Dale, T. Nelson. *The Commercial Marbles of Western Vermont.* United States Geological Survey Bulletin 521. Washington, DC: US Government Printing Office, 1912.

Dale, T. Nelson, Edwin C. Eckel, W. F. Hillebrand, and A. T. Coons. *Slate Deposits and Slate Industry of the United States.* United States Geological Survey Bulletin 275. Washington, DC: US Government Printing Office, 1906.

Eckert, Kathryn Bishop. *The Sandstone Architecture of the Lake Superior Region.* Detroit, MI: Wayne State University, 2000.

Georgia Marble Company. *Yesterday, Today, and Forever: The Story of Georgia Marble.* Tate, GA: Georgia Marble Company, n.d.

Gordon, Charles Henry. *The Marbles of Tennessee.* Nashville, TN: Tennessee State Geological Survey, 1911.

Hawes, George W., F. W. Sperr, and T. C. Kelly. *Report on the Building Stones of the United States, and Statistics of the Quarry Industry for 1880.* Washington, DC: US Census Office, 1883.

Heinrich, E. William. *Economic Geology of the Sand and Sandstone Resources of Michigan.* Lansing, MI: Michigan Department of Environmental Quality Geological Survey Division, 2001.

Hieb, James. *Sandstone Center of the World.* South Amherst, OH: Quarrytown.net Publishing, 2007.

Hinchey, Norman S. *Missouri Marble.* Report of Investigations, no. 3. Rolla, MO: Missouri State Geological Survey, 1946.

Howe, John Allen. *The Geology of Building Stones.* London: Routledge, 1910.

Indiana University. "Building a Nation: Indiana Limestone Photograph Collection." Accessed March 20, 2020. http://webapp1.dlib.indiana.edu/images/splash. htm?scope=images/VAC5094.

Jenkins, Joseph. *The Slate Roof Bible.* 3rd ed. Grove City, PA: Joseph C. Jenkins, 2016.

Lamar, J. E., and H. B. Willman. *Illinois Building Stones.* Report of Investigations 184. Urbana, IL: Division of the State Geological Survey, 1955.

Lent, Frank A. *Trade Names and Descriptions of Marbles, Limestones, Sandstones, Granites and Other Building Stones Quarried in the United States, Canada, and Other Countries.* New York: Stone Publishing, 1925.

McClymont, J. J. *A List of the World's Marbles.* Farmington, MI: Marble Institute of America, 1990.

McDonald, William H. *A Short History of Indiana Limestone.* Bedford, IN: Lawrence County Tourism Commission, 1995.

Merrill, George Perkins. *Stones for Building and Decoration.* 3rd ed. New York: John Wiley and Sons, 1908.

National Slate Association. *Slate Roofs.* Poultney, VT: National Slate Association, 2010.

Ohio Division of Geological Survey. "Ohio's Sandstone Industry." *Ohio Geology Newsletter,* Spring 1982.

Patton, John B., and Donald D. Carr. *The Salem Limestone in the Indiana Building-Stone District.* Occasional Paper 28. Bloomington, IN: Indiana Geological Survey, 1982.

Perazzo, Peggy B., and George Perazzo. Stone Quarries and Beyond (website). https://quarriesandbeyond.org.

Pivko, Daniel. "Natural Stones in Earth's History." *Acta Geologica Universitatis Comenianae* 58 (2003): 73–86.

Prouty, William Frederick. *Preliminary Report on the Crystalline and Other Marbles of Alabama.* Geological Survey of Alabama Bulletin 18. Tuscaloosa: University of Alabama, 1916.

Renwick, William George. *Marble and Marble Working.* New York: Van Nostrand, 1909.

Stone, Ralph. *Building Stones of Pennsylvania.* Bulletin M 15. Harrisburg, PA: Topographic and Geologic Survey, 1932.

Thiel, George A., and Carl E. Dutton. *The Architectural, Structural, and Monumental Stones of Minnesota.* Minneapolis: University of Minnesota Press, 1935.

Winkler, Erhard M. *Stone in Architecture: Properties, Durability.* 3rd ed. Berlin: Springer-Verlag, 1997.

OTHER BUILDING MATERIALS

Berry, George A., III, with Sharon S. Darling. *Common Clay: A History of American Terra Cotta Corporation, 1881–1966.* Crystal Lake, IL: TCR, 2003.

Buckley, Ernest Robertson. *The Clays and Clay Industries of Wisconsin.* Madison, WI: Wisconsin Geological and Natural History Survey, 1901.

Campbell, James W. P. *Brick: A World History.* London: Thames & Hudson, 2003.

Courland, Robert. *Concrete Planet.* Amherst, NY: Prometheus Books, 2011.

Gayle, Margot, David W. Look, and John G. Waite. *Metals in America's Historic Buildings.* Rev. ed. Washington, DC: National Park Service, 1992.

Peck, Herbert. *The Book of Rookwood Pottery.* New York: Crown, 1968.

Ries, Heinrich. *The Clays of Wisconsin and Their Uses.* Madison, WI: Wisconsin Geological and Natural History Survey, 1906.

Ries, Heinrich, and Henry Leighton. *History of the Clay-Working Industry in the United States.* New York: John Wiley & Sons, 1909.

ARCHITECTURE

Bach, Ira J., and Mary Lackritz Gray. *A Guide to Chicago's Public Sculpture.* Chicago: University of Chicago Press, 1983.

Bach, Ira J., and Susan Wolfson. *Chicago on Foot: Walking Tours of Chicago's Architecture.* 4th ed. Chicago: Chicago Review Press, 1987.

Bey, Lee. *Southern Exposure: The Overlooked Architecture of Chicago's South Side.* Evanston, IL: Northwestern University Press, 2019.

Blaser, Werner, ed. *Chicago Architecture: Holabird & Root, 1880–1992.* Basel: Birkhäuser, 1992.

Bluestone, Daniel. *Constructing Chicago.* New Haven, CT: Yale University Press, 1991.

Bruegmann, Robert. *The Architects and the City: Holabird & Roche of Chicago, 1880–1918.* Chicago: University of Chicago Press, 1997.

Bruegmann, Robert, ed. *Art Deco Chicago: Designing Modern America.* Chicago: Chicago Art Deco Society, 2018.

Chappell, Sally A. Kitt. *Architecture and Planning of Graham, Anderson, Probst and White, 1912–1936.* Chicago: University of Chicago Press, 1992.

City of Chicago. "Commission on Chicago Landmarks." https://www.chicago.gov/city/en/depts/dcd/supp_info/landmarks_commission.html.

Condit, Carl W. *Chicago 1910–23: Building, Planning and Urban Technology.* Chicago: University of Chicago Press, 1973.

Condit, Carl W. *The Chicago School of Architecture*. Chicago: University of Chicago Press, 1964.

Connors, Joseph. *The Robie House of Frank Lloyd Wright*. Chicago: University of Chicago Press, 1984.

Drury, John. *Old Chicago Houses*. Chicago: University of Chicago Press, 1941.

Hines, Thomas S. *Burnham of Chicago*. 2nd ed. Chicago: University of Chicago Press, 2009.

Hoffmann, Donald. *The Architecture of John Wellborn Root*. Chicago: University of Chicago Press, 1973.

Hoffmann, Donald. *Frank Lloyd Wright's Robie House: The Illustrated Story of an Architectural Masterpiece*. New York: Dover Publications, 1984.

Kamin, Blair. *Why Architecture Matters*. Chicago: University of Chicago Press, 2001.

Lane, George A. *Chicago Churches and Synagogues: An Architectural Pilgrimage*. Chicago: Loyola University Press, 1981.

Leslie, Thomas. *Chicago Skyscrapers 1871–1934*. Urbana: University of Illinois Press, 2013.

Monroe, Harriet. *John Wellborn Root: A Study of His Life and Work*. New York: Houghton Mifflin, 1912.

Morrison, Hugh. *Louis Sullivan: Prophet of Modern Architecture*. New York: W. W. Norton, 1935.

National Register of Historic Places Inventory. https://npgallery.nps.gov/NRHP/.

O'Gorman, James F. *Living Architecture: A Biography of H. H. Richardson*. New York: Simon & Schuster, 1997.

O'Gorman, James F. *Three American Architects*. Chicago: University of Chicago, 1991.

Sinkevitch, Alice, and Laurie McGovern Petersen, eds. *AIA Guide to Chicago*. 3rd ed. Urbana: University of Illinois Press, 2014.

Siry, Joseph. *Carson Pirie Scott: Louis Sullivan and the Chicago Department Store*. Chicago: University of Chicago Press, 1988.

Siry, Joseph. *The Chicago Auditorium Building*. Chicago: University of Chicago Press, 2002.

Stamper, John W. *Chicago's North Michigan Avenue*. Chicago: University of Chicago Press, 1991.

Twombly, Robert, and Narciso G. Menocal. *Louis Sullivan: The Poetry of Architecture*. New York: W. W. Norton, 2000.

Wolner, Edward D. *Henry Ives Cobb's Chicago*. Chicago: University of Chicago Press, 2011.

CIVIL ENGINEERING AND CONSTRUCTION

Hoffmann, Donald. "Pioneer Caisson Building Foundations: 1890." *Journal of the Society of Architectural Historians* 25, no. 1 (1966): 68–71.

Levy, Matthys, and Mario Salvadori. *Why Buildings Fall Down*. Rev. ed. New York: W. W. Norton, 2002.

Manierre, George. "A Description of Caisson Work to Bed Rock for Modern High Buildings." *Journal of the Illinois State Historical Society* 9, no. 3 (1916): 296–300.

Peck, Ralph B. *History of Building Foundations in Chicago*. Engineering Experiment Station Bulletin Series, no. 373. *University of Illinois Bulletin* 45, no. 29, January 2, 1948.

Randall, Frank A. *History of the Development of Building Construction in Chicago*. 2nd ed. Urbana: University of Illinois Press, 1999.

Schock, Robert E., and Eric J. Risberg. "120 Years of Caisson Foundations in Chicago." *International Conference on Case Histories in Geotechnical Engineering* 6, 2013. https://scholarsmine.mst.edu/icchge/7icchge/session10/6/.

CHICAGOLAND HISTORY

Andreas, A. T. *History of Chicago from the Earliest Period to the Present Time.* 3 vols. Chicago: A. T. Andreas, 1884.

Kallick, Sonia. *Lemont and Its People.* Louisville, KY: Chicago Spectrum Press, 1998.

Lanctot, Barbara. *A Walk through Graceland Cemetery.* Chicago: Chicago Architecture Foundation, 2011.

Land, John E. *Chicago, the Future Metropolis of the New World: Her Trade, Commerce and Industries.* Chicago: John E. Land, 1883.

Rand, McNally & Co.'s Bird's-Eye Views and Guide to Chicago. Chicago: Rand, McNally, 1893.

Schoon, Kenneth J. *Calumet Beginnings.* Bloomington: University of Indiana Press, 2003.

Smith, Carl. *Chicago's Great Fire.* New York: Atlantic Monthly Press, 2020.

Wilkie, Franc B. *Marquis' Hand-Book of Chicago.* Chicago: A. N. Marquis, 1885.

TRADE JOURNALS & NEWSPAPERS CONSULTED

American Architect and Building News

Architectural Forum

Architectural Record

Brick

Brickbuilder

Building Stone Magazine

Chicago Tribune

Daily Inter Ocean

Economist (Chicago)

Inland Architect and News Record

Manufacturer and Builder

Stone, an Illustrated Magazine

Stone World

Through the Ages

Western Architect

Index